All the Mathematics You Missed

Beginning graduate students in mathematics and other quantitat subjects are expected to have a daunting breadth of mathemati knowledge, but few have such a background. This book will h students see the broad outline of mathematics and to fill in the gaps their knowledge.

The author explains the basic points and a few key results of the m important undergraduate topics in mathematics, emphasizing t intuitions behind the subject. The topics include linear algebra, vect calculus, differential geometry, real analysis, point-set topolog differential equations, probability theory, complex analysis, abstr algebra, and more. An annotated bibliography offers a guide to furth reading and more rigorous foundations.

This book will be an essential resource for advanced undergradua and beginning graduate students in mathematics, the physical science engineering, computer science, statistics, and economics, and for anyo else who needs to quickly learn some serious mathematics.

Thomas A. Garrity is Professor of Mathematics at Williams College Williamstown, Massachusetts. He was an undergraduate at th University of Texas, Austin, and a graduate student at Brown Universit receiving his Ph.D. in 1986. From 1986 to 1989, he was G.C. Evan Instructor at Rice University. In 1989, he moved to Williams Colleg where he has been ever since except in 1992–3, when he spent the year the University of Washington, and 2000–1, when he spent the year at th University of Michigan, Ann Arbor.

All the Mathematics You Missed
But Need to Know for Graduate School

Thomas A. Garrity

Williams College

Figures by Lori Pedersen

PUBLISHED BY THE PRESS SYNDICATE OF THE UNIVERSITY
OF CAMBRIDGE
The Pitt Building, Trumpington Street, Cambridge, United Kingdom

CAMBRIDGE UNIVERSITY PRESS
The Edinburgh Building, Cambridge CB2 2RU, UK
40 West 20th Street, New York, NY 10011-4211, USA
10 Stamford Road, Oakleigh, VIC 3166, Australia
Ruiz de Alarcón 13, 28014 Madrid, Spain
Dock House, The Waterfront, Cape Town 8001, South Africa

http://www.cambridge.org

First published 2002

Printed in the United States of America

Typeface Palatino 10/12 pt.   *System* LATEX [AU]

A catalog record for this book is available from the British Library.

Library of Congress Cataloging in Publication Data

Garrity, Thomas A., 1959 –
  All the mathematics you missed : but need to know for graduate
  school / Thomas A. Garrity.
     p. cm.
  Includes bibliographical references and index.
  ISBN 0-521-79285-1 – ISBN 0-521-79707-1 (pb.)
  1. Mathematics. I. Title.

  QA37.3 .G37 2002
  510–dc21   2001037644

ISBN 0 521 79285 1 hardback
ISBN 0 521 79707 1 paperback

Dedicated to the Memory

of

Robert Mizner

Contents

Preface

Math is Exciting. We are living in the greatest age of mathematics ever seen. In the 1930s, there were some people who feared that the rising abstractions of the early twentieth century would either lead to mathematicians working on sterile, silly intellectual exercises or to mathematics splitting into sharply distinct subdisciplines, similar to the way natural philosophy split into physics, chemistry, biology and geology. But the very opposite has happened. Since World War II, it has become increasingly clear that mathematics is one unified discipline. What were separate areas now feed off of each other. Learning and creating mathematics is indeed a worthwhile way to spend one's life.

Math is Hard. Unfortunately, people are just not that good at mathematics. While intensely enjoyable, it also requires hard work and self-discipline. I know of no serious mathematician who finds math easy. In fact, most, after a few beers, will confess as to how stupid and slow they are. This is one of the personal hurdles that a beginning graduate student must face, namely how to deal with the profundity of mathematics in stark comparison to our own shallow understandings of mathematics. This is in part why the attrition rate in graduate school is so high. At the best schools, with the most successful retention rates, usually only about half of the people who start eventually get their PhDs. Even schools that are in the top twenty have at times had eighty percent of their incoming graduate students not finish. This is in spite of the fact that most beginning graduate students are, in comparison to the general population, amazingly good at mathematics. Most have found that math is one area in which they could shine. Suddenly, in graduate school, they are surrounded by people who are just as good (and who seem even better). To make matters worse, mathematics is a meritocracy. The faculty will not go out of their way to make beginning students feel good (this is not the faculty's job; their job is to discover new mathematics). The fact is that there are easier (though, for a mathematician, less satisfying) ways to make a living. There is truth in the statement that you must be driven to become a mathematician.

Mathematics is exciting, though. The frustrations should more than be compensated for by the thrills of learning and eventually creating (or discovering) new mathematics. That is, after all, the main goal for attending graduate school, to become a research mathematician. As with all creative endeavors, there will be emotional highs and lows. Only jobs that are routine and boring will not have these peaks and valleys. Part of the difficulty of graduate school is learning how to deal with the low times.

Goal of Book. The goal of this book is to give people at least a rough idea of the many topics that beginning graduate students at the best graduate schools are assumed to know. Since there is unfortunately far more that is needed to be known for graduate school and for research than it is possible to learn in a mere four years of college, few beginning students know all of these topics, but hopefully all will know at least some. Different people will know different topics. This strongly suggests the advantage of working with others.

There is another goal. Many nonmathematicians suddenly find that they need to know some serious math. The prospect of struggling with a text will legitimately seem for them to be daunting. Each chapter of this book will provide for these folks a place where they can get a rough idea and outline of the topic they are interested in.

As for general hints for helping sort out some mathematical field, certainly one should always, when faced with a new definition, try to find a simple example and a simple non-example. A non-example, by the way, is an example that almost, but not quite, satisfies the definition. But beyond finding these examples, one should examine the reason why the basic definitions were given. This leads to a split into two streams of thought for how to do mathematics. One can start with reasonable, if not naive, definitions and then prove theorems about these definitions. Frequently the statements of the theorems are complicated, with many different cases and conditions, and the proofs are quite convoluted, full of special tricks.

The other, more mid-twentieth century approach, is to spend quite a bit of time on the basic definitions, with the goal of having the resulting theorems be clearly stated and having straightforward proofs. Under this philosophy, any time there is a trick in a proof, it means more work needs to be done on the definitions. It also means that the definitions themselves take work to understand, even at the level of figuring out why anyone would care. But now the theorems can be cleanly stated and proved.

In this approach the role of examples becomes key. Usually there are basic examples whose properties are already known. These examples will shape the abstract definitions and theorems. The definitions in fact are made in order for the resulting theorems to give, for the examples, the answers we expect. Only then can the theorems be applied to new examples and cases whose properties are unknown.

For example, the correct notion of a derivative and thus of the slope of a tangent line is somewhat complicated. But whatever definition is chosen, the slope of a horizontal line (and hence the derivative of a constant function) must be zero. If the definition of a derivative does not yield that a horizontal line has zero slope, it is the definition that must be viewed as wrong, not the intuition behind the example.

For another example, consider the definition of the curvature of a plane curve, which is in Chapter Seven. The formulas are somewhat ungainly. But whatever the definitions, they must yield that a straight line has zero curvature, that at every point of a circle the curvature is the same and that the curvature of a circle with small radius must be greater than the curvature of a circle with a larger radius (reflecting the fact that it is easier to balance on the earth than on a basketball). If a definition of curvature does not do this, we would reject the definitions, not the examples.

Thus it pays to know the key examples. When trying to undo the technical maze of a new subject, knowing these examples will not only help explain why the theorems and definitions are what they are but will even help in predicting what the theorems must be.

Of course this is vague and ignores the fact that first proofs are almost always ugly and full of tricks, with the true insight usually hidden. But in learning the basic material, look for the key idea, the key theorem and then see how these shape the definitions.

Caveats for Critics. This book is far from a rigorous treatment of any topic. There is a deliberate looseness in style and rigor. I am trying to get the point across and to write in the way that most mathematicians talk to each other. The level of rigor in this book would be totally inappropriate in a research paper.

Consider that there are three tasks for any intellectual discipline:

1. Coming up with new ideas.

2. Verifying new ideas.

3. Communicating new ideas.

How people come up with new ideas in mathematics (or in any other field) is overall a mystery. There are at best a few heuristics in mathematics, such as asking if something is unique or if it is canonical. It is in verifying new ideas that mathematicians are supreme. Our standard is that there must be a rigorous proof. Nothing else will do. This is why the mathematical literature is so trustworthy (not that mistakes don't creep in, but they are usually not major errors). In fact, I would go as far as to say that if any discipline has as its standard of verification rigorous proof, than that discipline must be a part of mathematics. Certainly the main goal for a math major in the first few years of college is to learn what a rigorous proof is.

Unfortunately, we do a poor job of communicating mathematics. Every year there are millions of people who take math courses. A large number of people who you meet on the street or on the airplane have taken college level mathematics. How many enjoyed it? How many saw no real point to it? While this book is not addressed to that random airplane person, it is addressed to beginning graduate students, people who already enjoy mathematics but who all too frequently get blown out of the mathematical water by mathematics presented in an unmotivated, but rigorous, manner. There is no problem with being nonrigorous, as long as you know and clearly label when you are being nonrigorous.

Comments on the Bibliography. There are many topics in this book. While I would love to be able to say that I thoroughly know the literature on each of these topics, that would be a lie. The bibliography has been cobbled together from recommendations from colleagues, from books that I have taught from and books that I have used. I am confident that there are excellent texts that I do not know about. If you have a favorite, please let me know at tgarrity@williams.edu.

While this book was being written, Paulo Ney De Souza and Jorge-Nuno Silva wrote *Berkeley Problems in Mathematics* [26], which is an excellent collection of problems that have appeared over the years on qualifying exams (usually taken in the first or second year of graduate school) in the math department at Berkeley. In many ways, their book is the complement of this one, as their work is the place to go to when you want to test your computational skills while this book concentrates on underlying intuitions. For example, say you want to learn about complex analysis. You should first read chapter nine of this book to get an overview of the basics about complex analysis. Then choose a good complex analysis book and work most of its exercises. Then use the problems in De Souza and Silva as a final test of your knowledge.

Finally, the book *Mathematics, Form and Function* by Mac Lane [82], is excellent. It provides an overview of much of mathematics. I am listing it here because there was no other place where it could be naturally referenced. Second and third year graduate students should seriously consider reading this book.

Acknowledgments

First, I would like to thank Lori Pedersen for a wonderful job of creating the illustrations and diagrams for this book.

Many people have given feedback and ideas over the years. Nero Budar, Chris French and Richard Haynes were student readers of one of the early versions of this manuscript. Ed Dunne gave much needed advice and help. In the spring semester of 2000 at Williams, Tegan Cheslack-Postava, Ben Cooper and Ken Dennison went over the book line-by-line. Others who have given ideas have included Bill Lenhart, Frank Morgan, Cesar Silva, Colin Adams, Ed Burger, David Barrett, Sergey Fomin, Peter Hinman, Smadar Karni, Dick Canary, Jacek Miekisz, David James and Eric Schippers. During the final rush to finish this book, Trevor Arnold, Yann Bernard, Bill Correll, Jr., Bart Kastermans, Christopher Kennedy, Elizabeth Klodginski, Alex Köronya, Scott Kravitz, Steve Root and Craig Westerland have provided amazing help. Marissa Barschdorff texed a very early version of this manuscript. The Williams College Department of Mathematics and Statistics has been a wonderful place to write the bulk of this book; I thank all of my Williams' colleagues. The last revisions were done while I have been on sabbatical at the University of Michigan, another great place to do mathematics. I would like to thank my editor at Cambridge, Lauren Cowles, and also Caitlin Doggart at Cambridge. Gary Knapp has throughout provided moral support and gave a close, detailed reading to an early version of the manuscript. My wife, Lori, has also given much needed encouragement and has spent many hours catching many of my mistakes. To all I owe thanks.

Finally, near the completion of this work, Bob Mizner passed away at an early age. It is in his memory that I dedicate this book (though no doubt he would have disagreed with most of my presentations and choices of topics; he definitely would have made fun of the lack of rigor).

On the Structure of Mathematics

If you look at articles in current journals, the range of topics seems immense. How could anyone even begin to make sense out of all of these topics? And indeed there is a glimmer of truth in this. People cannot effortlessly switch from one research field to another. But not all is chaos. There are at least two ways of placing some type of structure on all of mathematics.

Equivalence Problems

Mathematicians want to know when things are the same, or, when they are equivalent. What is meant by the same is what distinguishes one branch of mathematics from another. For example, a topologist will consider two geometric objects (technically, two topological spaces) to be the same if one can be twisted and bent, but not ripped, into the other. Thus for a topologist, we have

$$\bigcirc = \bigcirc = \square$$

To a differential topologist, two geometric objects are the same if one can be smoothly bent and twisted into the other. By smooth we mean that no sharp edges can be introduced. Then

$$\bigcirc = \bigcirc \neq \square$$

The four sharp corners of the square are what prevent it from being equivalent to the circle.

For a differential geometer, the notion of equivalence is even more restrictive. Here two objects are the same not only if one can be smoothly bent and twisted into the other but also if the curvatures agree. Thus for the differential geometer, the circle is no longer equivalent to the ellipse:

As a first pass to placing structure on mathematics, we can view an area of mathematics as consisting of certain *Objects*, coupled with the notion of *Equivalence* between these objects. We can explain equivalence by looking at the allowed *Maps*, or functions, between the objects. At the beginning of most chapters, we will list the Objects and the Maps between the objects that are key for that subject. The *Equivalence Problem* is of course the problem of determining when two objects are the same, using the allowable maps.

If the equivalence problem is easy to solve for some class of objects, then the corresponding branch of mathematics will no longer be active. If the equivalence problem is too hard to solve, with no known ways of attacking the problem, then the corresponding branch of mathematics will again not be active, though of course for opposite reasons. The hot areas of mathematics are precisely those for which there are rich partial but not complete answers to the equivalence problem. But what could we mean by a partial answer?

Here enters the notion of invariance. Start with an example. Certainly the circle, as a topological space, is different from two circles,

since a circle has only one connected component and two circles have two connected components. We map each topological space to a positive integer, namely the number of connected components of the topological space. Thus we have:

$$\text{Topological Spaces} \rightarrow \text{Positive Integers}.$$

The key is that the number of connected components for a space cannot change under the notion of topological equivalence (under bendings and twistings). We say that the number of connected components is an *invariant* of a topological space. Thus if the spaces map to different numbers, meaning that they have different numbers of connected components, then the two spaces cannot be topologically equivalent.

Of course, two spaces can have the same number of connected components and still be different. For example, both the circle and the sphere

have only one connected component, but they are different. (These can be distinguished by looking at each space's dimension, which is another topological invariant.) The goal of topology is to find enough invariants to be able to always determine when two spaces are different or the same. This has not come close to being done. Much of algebraic topology maps each space not to invariant numbers but to other types of algebraic objects, such as groups and rings. Similar techniques show up throughout mathematics. This provides for tremendous interplay between different branches of mathematics.

The Study of Functions

The mantra that we should all chant each night before bed is:

> Functions describe the World.

To a large extent what makes mathematics so useful to the world is that seemingly disparate real-world situations can be described by the same type of function. For example, think of how many different problems can be recast as finding the maximum or minimum of a function.

Different areas of mathematics study different types of functions. Calculus studies differentiable functions from the real numbers to the real numbers, algebra studies polynomials of degree one and two (in high school) and permutations (in college), linear algebra studies linear functions, or matrix multiplication.

Thus in learning a new area of mathematics, you should always "find the function" of interest. Hence at the beginning of most chapters we will state the type of function that will be studied.

Equivalence Problems in Physics

Physics is an experimental science. Hence any question in physics must eventually be answered by performing an experiment. But experiments come down to making observations, which usually are described by certain computable numbers, such as velocity, mass or charge. Thus the experiments in physics are described by numbers that are read off in the lab. More succinctly, physics is ultimately:

$$\boxed{\text{Numbers in Boxes}}$$

where the boxes are various pieces of lab machinery used to make measurements. But different boxes (different lab set-ups) can yield different numbers, even if the underlying physics is the same. This happens even at the trivial level of choice of units.

More deeply, suppose you are modeling the physical state of a system as the solution of a differential equation. To write down the differential equation, a coordinate system must be chosen. The allowed changes of coordinates are determined by the physics. For example, Newtonian physics can be distinguished from Special Relativity in that each has different allowable changes of coordinates.

Thus while physics is 'Numbers in Boxes', the true questions come down to when different numbers represent the same physics. But this is an equivalence problem; mathematics comes to the fore. (This explains in part the heavy need for advanced mathematics in physics.) Physicists want to find physics invariants. Usually, though, physicists call their invariants 'Conservation Laws'. For example, in classical physics the conservation of energy can be recast as the statement that the function that represents energy is an invariant function.

Brief Summaries of Topics

0.1 Linear Algebra

Linear algebra studies linear transformations and vector spaces, or in another language, matrix multiplication and the vector space $\mathbf{R}^n$. You should know how to translate between the language of abstract vector spaces and the language of matrices. In particular, given a basis for a vector space, you should know how to represent any linear transformation as a matrix. Further, given two matrices, you should know how to determine if these matrices actually represent the same linear transformation, but under different choices of bases. The key theorem of linear algebra is a statement that gives many equivalent descriptions for when a matrix is invertible. These equivalences should be known cold. You should also know why eigenvectors and eigenvalues occur naturally in linear algebra.

0.2 Real Analysis

The basic definitions of a limit, continuity, differentiation and integration should be known and understood in terms of ϵ's and δ's. Using this ϵ and δ language, you should be comfortable with the idea of uniform convergence of functions.

0.3 Differentiating Vector-Valued Functions

The goal of the Inverse Function Theorem is to show that a differentiable function $f : \mathbf{R}^n \to \mathbf{R}^n$ is locally invertible if and only if the determinant of its derivative (the Jacobian) is non-zero. You should be comfortable with what it means for a vector-valued function to be differentiable, why its derivative must be a linear map (and hence representable as a matrix, the Jacobian) and how to compute the Jacobian. Further, you should know the statement of the Implicit Function Theorem and see why is is closely related to the Inverse Function Theorem.

0.4   Point Set Topology

You should understand how to define a topology in terms of open sets and how to express the idea of continuous functions in terms of open sets. The standard topology on $\mathbf{R}^n$ must be well understood, at least to the level of the Heine-Borel Theorem. Finally, you should know what a metric space is and how a metric can be used to define open sets and hence a topology.

0.5   Classical Stokes' Theorems

You should know about the calculus of vector fields. In particular, you should know how to compute, and know the geometric interpretations behind, the curl and the divergence of a vector field, the gradient of a function and the path integral along a curve. Then you should know the classical extensions of the Fundamental Theorem of Calculus, namely the Divergence Theorem and Stokes' Theorem. You should especially understand why these are indeed generalizations of the Fundamental Theorem of Calculus.

0.6   Differential Forms and Stokes' Theorem

Manifolds are naturally occurring geometric objects. Differential k-forms are the tools for doing calculus on manifolds. You should know the various ways for defining a manifold, how to define and to think about differential k-forms, and how to take the exterior derivative of a k-form. You should also be able to translate from the language of k-forms and exterior derivatives to the language from Chapter Five on vector fields, gradients, curls and divergences. Finally, you should know the statement of Stokes' Theorem, understand why it is a sharp quantitative statement about the equality of the integral of a k-form on the boundary of a $(k+1)$-dimensional manifold with the integral of the exterior derivative of the k-form on the manifold, and how this Stokes' Theorem has as special cases the Divergence Theorem and the Stokes' Theorem from the previous chapter.

0.7   Curvature for Curves and Surfaces

Curvature, in all of its manifestations, attempts to measure the rate of change of the directions of tangent spaces of geometric objects. You should

know how to compute the curvature of a plane curve, the curvature and
the torsion of a space curve and the two principal curvatures, in terms of
the Hessian, of a surface in space.

0.8   Geometry

Different geometries are built out of different axiomatic systems. Given a
line l and a point p not on l, Euclidean geometry assumes that there is
exactly one line containing p parallel to l, hyperbolic geometry assumes
that there is more than one line containing p parallel to l, and elliptic
geometries assume that there is no line parallel to l. You should know
models for hyperbolic geometry, single elliptic geometry and double elliptic
geometry. Finally, you should understand why the existence of such models
implies that all of these geometries are mutually consistent.

0.9   Complex Analysis

The main point is to recognize and understand the many equivalent ways
for describing when a function can be analytic. Here we are concerned with
functions $f : U \rightarrow \mathbf{C}$, where U is an open set in the complex numbers
$\mathbf{C}$. You should know that such a function $f(z)$ is said to be *analytic* if it
satisfies any of the following equivalent conditions:
a) For all $z_0 \in U$,

$$\lim_{z \to z_0} \frac{f(z) - f(z_0)}{z - z_0}$$

exists.
b)The real and imaginary parts of the function f satisfy the *Cauchy-
Riemann equations*:

$$\frac{\partial Ref}{\partial x} = \frac{\partial Imf}{\partial y}$$

and

$$\frac{\partial Ref}{\partial y} = -\frac{\partial Imf}{\partial x}.$$

c) If γ is any counterclockwise simple loop in $\mathbf{C} = \mathbf{R}^2$ and if z_0 is any complex
number in the interior of γ, then

$$f(z_0) = \frac{1}{2\pi i} \int_{\gamma} \frac{f(z)}{z - z_0} dz.$$

This is the *Cauchy Integral formula*.

d) For any complex number z_0, there is an open neighborhood in $\mathbf{C} = \mathbf{R}^2$ of z_0 on which

$$f(z) = \sum_{k=o}^{\infty} a_k (z - z_0)^k,$$

is a uniformly converging series.

Further, if $f : U \to \mathbf{C}$ is analytic and if $f'(z_0) \neq 0$, then at z_0, the function f is conformal (i.e., angle-preserving), viewed as a map from $\mathbf{R}^2$ to $\mathbf{R}^2$.

0.10 Countability and the Axiom of Choice

You should know what it means for a set to be countably infinite. In particular, you should know that the integers and rationals are countably infinite while the real numbers are uncountably infinite. The statement of the Axiom of Choice and the fact that it has many seemingly bizarre equivalences should also be known.

0.11 Algebra

Groups, the basic object of study in abstract algebra, are the algebraic interpretations of geometric symmetries. One should know the basics about groups (at least to the level of the Sylow Theorem, which is a key tool for understanding finite groups), rings and fields. You should also know Galois Theory, which provides the link between finite groups and the finding of the roots of a polynomial and hence shows the connections between high school and abstract algebra. Finally, you should know the basics behind representation theory, which is how one relates abstract groups to groups of matrices.

0.12 Lebesgue Integration

You should know the basic ideas behind Lebesgue measure and integration, at least to the level of the Lebesgue Dominating Convergence Theorem, and the concept of sets of measure zero.

0.13 Fourier Analysis

You should know how to find the Fourier series of a periodic function, the Fourier integral of a function, the Fourier transform, and how Fourier series relate to Hilbert spaces. Further, you should see how Fourier transforms can be used to simplify differential equations.

0.14 Differential Equations

Much of physics, economics, mathematics and other sciences comes down to trying to find solutions to differential equations. One should know that the goal in differential equations is to find an unknown function satisfying an equation involving derivatives. Subject to mild restrictions, there are always solutions to ordinary differential equations. This is most definitely not the case for partial differential equations, where even the existence of solutions is frequently unknown. You should also be familiar with the three traditional classes of partial differential equations: the heat equation, the wave equation and the Laplacian.

0.15 Combinatorics and Probability Theory

Both elementary combinatorics and basic probability theory reduce to problems in counting. You should know that

$$\binom{n}{k} = \frac{n!}{k!(n-k)!}$$

is the number of ways of choosing k elements from n elements. The relation of $\binom{n}{k}$ to the binomial theorem for polynomials is useful to have handy for many computations. Basic probability theory should be understood. In particular one should understand the terms: sample space, random variable (both its intuitions and its definition as a function), expected value and variance. One should definitely understand why counting arguments are critical for calculating probabilities of finite sample spaces. The link between probability and integral calculus can be seen in the various versions of the Central Limit Theorem, the ideas of which should be known.

0.16 Algorithms

You should understand what is meant by the complexity of an algorithm, at least to the level of understanding the question P=NP. Basic graph theory should be known; for example, you should see why a tree is a natural structure for understanding many algorithms. Numerical Analysis is the study of algorithms for approximating the answer to computations in mathematics. As an example, you should understand Newton's method for approximating the roots of a polynomial.

Chapter 1

Linear Algebra

Basic Object:	Vector Spaces
Basic Map:	Linear Transformations
Basic Goal:	Equivalences for the Invertibility of Matrices

1.1 Introduction

Though a bit of an exaggeration, it can be said that a mathematical problem can be solved only if it can be reduced to a calculation in linear algebra. And a calculation in linear algebra will reduce ultimately to the solving of a system of linear equations, which in turn comes down to the manipulation of matrices. Throughout this text and, more importantly, throughout mathematics, linear algebra is a key tool (or more accurately, a collection of intertwining tools) that is critical for doing calculations.

The power of linear algebra lies not only in our ability to manipulate matrices in order to solve systems of linear equations. The abstraction of these concrete objects to the ideas of vector spaces and linear transformations allows us to see the common conceptual links between many seemingly disparate subjects. (Of course, this is the advantage of any good abstraction.) For example, the study of solutions to linear differential equations has, in part, the same feel as trying to model the hood of a car with cubic polynomials, since both the space of solutions to a linear differential equation and the space of cubic polynomials that model a car hood form vector spaces.

The key theorem of linear algebra, discussed in section six, gives many equivalent ways of telling when a system of n linear equations in n unknowns has a solution. Each of the equivalent conditions is important. What is remarkable and what gives linear algebra its oomph is that they are all the same.

1.2   The Basic Vector Space $\mathbf{R}^n$

The quintessential vector space is $\mathbf{R}^n$, the set of all n-tuples of real numbers

$$\{(x_1, \ldots, x_n) : x_i \in \mathbf{R}\}.$$

As we will see in the next section, what makes this a vector space is that we can add together two n-tuples to get another n-tuple:

$$(x_1, \ldots, x_n) + (y_1, \ldots, y_n) = (x_1 + y_1, \ldots, x_n + y_n)$$

and that we can multiply each n-tuple by a real number λ:

$$\lambda(x_1, \ldots, x_n) = (\lambda x_1, \ldots, \lambda x_n)$$

to get another n-tuple. Of course each n-tuple is usually called a vector and the real numbers λ are called scalars. When $n = 2$ and when $n = 3$ all of this reduces to the vectors in the plane and in space that most of us learned in high school.

The natural map from some $\mathbf{R}^n$ to an $\mathbf{R}^m$ is given by matrix multiplication. Write a vector $\mathbf{x} \in \mathbf{R}^n$ as a column vector:

$$\mathbf{x} = \begin{pmatrix} x_1 \\ \vdots \\ x_n \end{pmatrix}.$$

Similarly, we can write a vector in $\mathbf{R}^m$ as a column vector with m entries. Let A be an $m \times n$ matrix

$$A = \begin{pmatrix} a_{11} & a_{12} & \cdots & a_{1n} \\ \vdots & \vdots & \vdots & \vdots \\ a_{m1} & \cdots & \cdots & a_{mn} \end{pmatrix}.$$

Then $A\mathbf{x}$ is the m-tuple:

$$A\mathbf{x} = \begin{pmatrix} a_{11} & a_{12} & \cdots & a_{1n} \\ \vdots & \vdots & \vdots & \vdots \\ a_{m1} & \cdots & \cdots & a_{mn} \end{pmatrix} \begin{pmatrix} x_1 \\ \vdots \\ x_n \end{pmatrix} = \begin{pmatrix} a_{11}x_1 + \ldots + a_{1n}x_n \\ \vdots \\ a_{m1}x_1 + \ldots + a_{mn}x_n \end{pmatrix}.$$

For any two vectors $\mathbf{x}$ and $\mathbf{y}$ in $\mathbf{R}^n$ and any two scalars λ and μ, we have

$$A(\lambda\mathbf{x} + \mu\mathbf{y}) = \lambda A\mathbf{x} + \mu A\mathbf{y}.$$

In the next section we will use the linearity of matrix multiplication to motivate the definition for a linear transformation between vector spaces.

Now to relate all of this to the solving of a system of linear equations. Suppose we are given numbers $b_1, \ldots, b_m$ and numbers $a_{11}, \ldots, a_{mn}$. Our goal is to find n numbers $x_1, \ldots, x_n$ that solve the following system of linear equations:

$$
\begin{aligned}
a_{11}x_1 + \cdots + a_{1n}x_n &= b_1 \\
&\vdots \\
a_{m1}x_1 + \cdots + a_{mn}x_n &= b_m.
\end{aligned}
$$

Calculations in linear algebra will frequently reduce to solving a system of linear equations. When there are only a few equations, we can find the solutions by hand, but as the number of equations increases, the calculations quickly turn from enjoyable algebraic manipulations into nightmares of notation. These nightmarish complications arise not from any single theoretical difficulty but instead stem solely from trying to keep track of the many individual minor details. In other words, it is a problem in bookkeeping.

Write

$$
\mathbf{b} = \begin{pmatrix} b_1 \\ \vdots \\ b_m \end{pmatrix}, \quad A = \begin{pmatrix} a_{11} & a_{12} & \cdots & a_{1n} \\ \vdots & \vdots & \vdots & \vdots \\ a_{m1} & \cdots & \cdots & a_{mn} \end{pmatrix}
$$

and our unknowns as

$$
\mathbf{x} = \begin{pmatrix} x_1 \\ \vdots \\ x_n \end{pmatrix}.
$$

Then we can rewrite our system of linear equations in the more visually appealing form of

$$
A\mathbf{x} = \mathbf{b}.
$$

When $m > n$ (when there are more equations than unknowns), we expect there to be, in general, no solutions. For example, when $m = 3$ and $n = 2$, this corresponds geometrically to the fact that three lines in a plane will usually have no common point of intersection. When $m < n$ (when there are more unknowns than equations), we expect there to be, in general, many solutions. In the case when $m = 2$ and $n = 3$, this corresponds geometrically to the fact that two planes in space will usually intersect in an entire line. Much of the machinery of linear algebra deals with the remaining case when $m = n$.

Thus we want to find the $n \times 1$ column vector $\mathbf{x}$ that solves $A\mathbf{x} = \mathbf{b}$, where A is a given $n \times n$ matrix and $\mathbf{b}$ is a given $n \times 1$ column vector.

Suppose that the square matrix A has an inverse matrix A^{-1} (which means that A^{-1} is also $n \times n$ and more importantly that $A^{-1}A = I$, with I the identity matrix). Then our solution will be

$$\mathbf{x} = A^{-1}\mathbf{b}$$

since

$$A\mathbf{x} = A(A^{-1}\mathbf{b}) = I\mathbf{b} = \mathbf{b}.$$

Thus solving our system of linear equations comes down to understanding when the $n \times n$ matrix A has an inverse. (If an inverse matrix exists, then there are algorithms for its calculations.)

The key theorem of linear algebra, stated in section six, is in essence a list of many equivalences for when an $n \times n$ matrix has an inverse and is thus essential to understanding when a system of linear equations can be solved.

1.3   Vector Spaces and Linear Transformations

The abstract approach to studying systems of linear equations starts with the notion of a vector space.

Definition 1.3.1 *A set V is a* vector space *over the real numbers*[1] **R** *if there are maps:*

1. $\mathbf{R} \times V \to V$, *denoted by $a \cdot v$ or av for all real numbers a and elements v in V,*

2. $V \times V \to V$, *denoted by $v + w$ for all elements v and w in the vector space V,*

with the following properties:
   a) There is an element 0, in V such that $0 + v = v$ for all $v \in V$.
   b) For each $v \in V$, there is an element $(-v) \in V$ with $v + (-v) = 0$.
   c) For all $v, w \in V$, $v + w = w + v$.
   d) For all $a \in \mathbf{R}$ and for all $v, w \in V$, we have that $a(v + w) = av + aw$.
   e) For all $a, b \in \mathbf{R}$ and all $v \in V$, $a(bv) = (a \cdot b)v$.
   f) For all $a, b \in \mathbf{R}$ and all $v \in V$, $(a + b)v = av + bv$.
   g) For all $v \in V$, $1 \cdot v = v$.

---

[1]The real numbers can be replaced by the complex numbers and in fact by any field (which will be defined in Chapter Eleven on algebra).

As a matter of notation, and to agree with common usage, the elements of a vector space are called *vectors* and the elements of **R** (or whatever field is being used) *scalars*. Note that the space $\mathbf{R}^n$ given in the last section certainly satisfies these conditions.

The natural map between vector spaces is that of a linear transformation.

Definition 1.3.2 *A linear transformation* $T : V \to W$ *is a function from a vector space* V *to a vector space* W *such that for any real numbers* a_1 *and* a_2 *and any vectors* v_1 *and* v_2 *in* V, *we have*

$$T(a_1 v_1 + a_2 v_2) = a_1 T(v_1) + a_2 T(v_2).$$

Matrix multiplication from an $\mathbf{R}^n$ to an $\mathbf{R}^m$ gives an example of a linear transformation.

Definition 1.3.3 *A subset* U *of a vector space* V *is a* subspace *of* V *if* U *is itself a vector space.*

In practice, it is usually easy to see if a subset of a vector space is in fact a subspace, by the following proposition, whose proof is left to the reader:

Proposition 1.3.1 *A subset* U *of a vector space* V *is a subspace of* V *if* U *is closed under addition and scalar multiplication.*

Given a linear transformation $T : V \to W$, there are naturally occurring subspaces of both V and W.

Definition 1.3.4 *If* $T : V \to W$ *is a linear transformation, then the* kernel *of* T *is:*

$$ker(T) = \{v \in V : T(v) = 0\}$$

and the image *of* T *is*

$$Im(T) = \{w \in W : \text{ there exists a } v \in V \text{ with } T(v) = w\}.$$

The kernel is a subspace of V, since if v_1 and v_2 are two vectors in the kernel and if a and b are any two real numbers, then

$$
\begin{aligned}
T(a v_1 + b v_2) &= aT(v_1) + bT(v_2) \\
&= a \cdot 0 + b \cdot 0 \\
&= 0.
\end{aligned}
$$

In a similar way we can show that the image of T is a subspace of W.

If the only vector spaces that ever occurred were column vectors in $\mathbf{R}^n$, then even this mild level of abstraction would be silly. This is not the case.

Here we look at only one example. Let $C^k[0,1]$ be the set of all real-valued functions with domain the unit interval $[0,1]$:

$$f : [0,1] \to \mathbf{R}$$

such that the kth derivative of f exists and is continuous. Since the sum of any two such functions and a multiple of any such function by a scalar will still be in $C^k[0,1]$, we have a vector space. Though we will officially define dimension next section, $C^k[0,1]$ will be infinite dimensional (and thus definitely not some $\mathbf{R}^n$). We can view the derivative as a linear transformation from $C^k[0,1]$ to those functions with one less derivative, $C^{k-1}[0,1]$:

$$\frac{\mathrm{d}}{\mathrm{d}x} : C^k[0,1] \to C^{k-1}[0,1].$$

The kernel of $\frac{\mathrm{d}}{\mathrm{d}x}$ consists of those functions with $\frac{\mathrm{d}f}{\mathrm{d}x} = 0$, namely constant functions.

Now consider the differential equation

$$\frac{\mathrm{d}^2 f}{\mathrm{d}x^2} + 3\frac{\mathrm{d}f}{\mathrm{d}x} + 2f = 0.$$

Let T be the linear transformation:

$$T = \frac{\mathrm{d}^2}{\mathrm{d}x^2} + 3\frac{\mathrm{d}}{\mathrm{d}x} + 2I : C^2[0,1] \to C^0[0,1].$$

The problem of finding a solution $f(x)$ to the original differential equation can now be translated to finding an element of the kernel of T. This suggests the possibility (which indeed is true) that the language of linear algebra can be used to understand solutions to (linear) differential equations.

1.4   Bases, Dimension, and Linear Transformations as Matrices

Our next goal is to define the dimension of a vector space.

Definition 1.4.1 *A set of vectors $(v_1, \ldots, v_n)$ form a* basis *for the vector space V if given any vector v in V, there are unique scalars $a_1, \ldots, a_n \in \mathbf{R}$ with $v = a_1 v_1 + \ldots + a_n v_n$.*

Definition 1.4.2 *The* dimension *of a vector space V, denoted by $dim(V)$, is the number of elements in a basis.*

As it is far from obvious that the number of elements in a basis will always be the same, no matter which basis is chosen, in order to make the definition of the dimension of a vector space well-defined we need the following theorem (which we will not prove):

Theorem 1.4.1 *All bases of a vector space V have the same number of elements.*

For $\mathbf{R}^n$, the usual basis is

$$\{(1, 0, ..., 0), (0, 1, 0, ..., 0), ..., (0, ..., 0, 1)\}.$$

Thus $\mathbf{R}^n$ is n dimensional. Of course if this were not true, the above definition of dimension would be wrong and we would need another. This is an example of the principle mentioned in the introduction. We have a good intuitive understanding of what dimension should mean for certain specific examples: a line needs to be one dimensional, a plane two dimensional and space three dimensional. We then come up with a sharp definition. If this definition gives the "correct" answer for our three already understood examples, we are somewhat confident that the definition has indeed captured what is meant by, in this case, dimension. Then we can apply the definition to examples where our intuitions fail.

Linked to the idea of a basis is:

Definition 1.4.3 *Vectors $(v_1, \ldots, v_n)$ in a vector space V are* linearly independent *if whenever*

$$a_1 v_1 + \cdots + a_n v_n = 0,$$

it must be the case that the scalars $a_1, \ldots, a_n$ must all be zero.

Intuitively, a collection of vectors are linearly independent if they all point in different directions. A basis consists then in a collection of linearly independent vectors that span the vector space, where by span we mean:

Definition 1.4.4 *A set of vectors $(v_1, \ldots, v_n)$* span *the vector space V if given any vector v in V, there are scalars $a_1, \ldots, a_n \in \mathbf{R}$ with $v = a_1 v_1 + \cdots + a_n v_n$.*

Our goal now is to show how all linear transformations $T : V \to W$ between finite-dimensional spaces can be represented as matrix multiplication, provided we fix bases for the vector spaces V and W.

First fix a basis $\{v_1, ..., v_n\}$ for V and a basis $\{w_1, ..., w_m\}$ for W. Before looking at the linear transformation T, we need to show how each element of the n-dimensional space V can be represented as a column vector in $\mathbf{R}^n$ and how each element of the m-dimensional space W can be represented as a column vector of $\mathbf{R}^m$. Given any vector v in V, by the definition of basis, there are unique real numbers $a_1, ..., a_n$ with

$$v = a_1 v_1 + \cdots + a_n v_n.$$

We thus represent the vector v with the column vector:

$$\begin{pmatrix} a_1 \\ \vdots \\ a_n \end{pmatrix}.$$

Similarly, for any vector w in W, there are unique real numbers $b_1, ..., b_m$ with

$$w = b_1 w_1 + \cdots + b_m w_m.$$

Here we represent w as the column vector

$$\begin{pmatrix} b_1 \\ \vdots \\ b_m \end{pmatrix}.$$

Note that we have established a correspondence between vectors in V and W and column vectors $\mathbf{R}^n$ and $\mathbf{R}^m$, respectively. More technically, we can show that V is isomorphic to $\mathbf{R}^n$ (meaning that there is a one-one, onto linear transformation from V to $\mathbf{R}^n$) and that W is isomorphic to $\mathbf{R}^m$, though it must be emphasized that the actual correspondence only exists after a basis has been chosen (which means that while the isomorphism exists, it is not canonical; this is actually a big deal, as in practice it is unfortunately often the case that no basis is given to us).

We now want to represent a linear transformation $T : V \to W$ as an $m \times n$ matrix A. For each basis vector v_i in the vector space V, $T(v_i)$ will be a vector in W. Thus there will exist real numbers $a_{1i}, \ldots, a_{mi}$ such that

$$T(v_i) = a_{1i} w_1 + \cdots + a_{mi} w_m.$$

We want to see that the linear transformation T will correspond to the $m \times n$ matrix

$$A = \begin{pmatrix} a_{11} & a_{12} & \cdots & a_{1n} \\ \vdots & \vdots & \vdots & \vdots \\ a_{m1} & \cdots & \cdots & a_{mn} \end{pmatrix}.$$

Given any vector v in V, with $v = a_1 v_1 + \cdots + a_n v_n$, we have

$$\begin{aligned} T(v) &= T(a_1 v_1 + \cdots + a_n v_n) \\ &= a_1 T(v_1) + \cdots + a_n T(v_n) \\ &= a_1 (a_{11} w_1 + \cdots + a_{m1} w_m) + \cdots \\ &\quad + a_n (a_{1n} w_1 + \cdots + a_{mn} w_m). \end{aligned}$$

But under the correspondences of the vector spaces with the various column spaces, this can be seen to correspond to the matrix multiplication of A times the column vector corresponding to the vector v:

$$\begin{pmatrix} a_{11} & a_{12} & \cdots & a_{1n} \\ \vdots & \vdots & \vdots & \vdots \\ a_{m1} & \cdots & \cdots & a_{mn} \end{pmatrix} \begin{pmatrix} a_1 \\ \vdots \\ a_n \end{pmatrix} = \begin{pmatrix} b_1 \\ \vdots \\ b_m \end{pmatrix}.$$

Note that if $T : V \to V$ is a linear transformation from a vector space to itself, then the corresponding matrix will be $n \times n$, a square matrix.

Given different bases for the vector spaces V and W, the matrix associated to the linear transformation T will change. A natural problem is to determine when two matrices actually represent the same linear transformation, but under different bases. This will be the goal of section seven.

1.5 The Determinant

Our next task is to give a definition for the determinant of a matrix. In fact, we will give three alternative descriptions of the determinant. All three are equivalent; each has its own advantages.

Our first method is to define the determinant of a 1×1 matrix and then to define recursively the determinant of an $n \times n$ matrix.

Since 1×1 matrices are just numbers, the following should not at all be surprising:

Definition 1.5.1 *The determinant of a 1×1 matrix (a) is the real-valued function*

$$\det(a) = a.$$

This should not yet seem significant.

Before giving the definition of the determinant for a general $n \times n$ matrix, we need a little notation. For an $n \times n$ matrix

$$A = \begin{pmatrix} a_{11} & a_{12} & \cdots & a_{1n} \\ \vdots & \vdots & \vdots & \vdots \\ a_{n1} & \cdots & \cdots & a_{nn} \end{pmatrix},$$

denote by A_{ij} the $(n-1) \times (n-1)$ matrix obtained from A by deleting the ith row and the jth column. For example, if $A = \begin{pmatrix} a_{11} & a_{12} \\ a_{21} & a_{22} \end{pmatrix}$, then $A_{12} = (a_{21})$. Similarly if $A = \begin{pmatrix} 2 & 3 & 5 \\ 6 & 4 & 9 \\ 7 & 1 & 8 \end{pmatrix}$, then $A_{12} = \begin{pmatrix} 6 & 9 \\ 7 & 8 \end{pmatrix}$.

Since we have a definition for the determinant for 1×1 matrices, we will now assume by induction that we know the determinant of any $(n-1) \times (n-1)$ matrix and use this to find the determinant of an $n \times n$ matrix.

Definition 1.5.2 *Let A be an $n \times n$ matrix. Then the* determinant *of A is*

$$\det(A) = \sum_{k=1}^{n} (-1)^{k+1} a_{1k} \det(A_{1k}).$$

Thus for $A = \begin{pmatrix} a_{11} & a_{12} \\ a_{21} & a_{22} \end{pmatrix}$, we have

$$\det(A) = a_{11} \det(A_{11}) - a_{12} \det(A_{12}) = a_{11}a_{22} - a_{12}a_{21},$$

which is what most of us think of as the determinant. The determinant of our above 3×3 matrix is:

$$\det \begin{pmatrix} 2 & 3 & 5 \\ 6 & 4 & 9 \\ 7 & 1 & 8 \end{pmatrix} = 2 \det \begin{pmatrix} 4 & 9 \\ 1 & 8 \end{pmatrix} - 3 \det \begin{pmatrix} 6 & 9 \\ 7 & 8 \end{pmatrix} + 5 \det \begin{pmatrix} 6 & 4 \\ 7 & 1 \end{pmatrix}.$$

While this definition is indeed an efficient means to describe the determinant, it obscures most of the determinant's uses and intuitions.

The second way we can describe the determinant has built into it the key algebraic properties of the determinant. It highlights function-theoretic properties of the determinant.

Denote the $n \times n$ matrix A as $A = (A_1, ..., A_n)$, where A_i denotes the ith column:

$$A_i = \begin{pmatrix} a_{1i} \\ a_{2i} \\ \vdots \\ a_{ni} \end{pmatrix}.$$

Definition 1.5.3 *The* determinant *of A is defined as the unique real-valued function*

$$\det : Matrices \to \mathbf{R}$$

satisfying:
a) $\det(A_1, ..., \lambda A_k, ..., A_n) = \lambda \det(A_1, ..., A_k)$.
b) $\det(A_1, ..., A_k + \lambda A_i, ..., A_n) = \det(A_1, ..., A_n)$ *for $k \neq i$.*
c) $\det$ *(Identity matrix)* $= 1$.

Thus, treating each column vector of a matrix as a vector in $\mathbf{R}^n$, the determinant can be viewed as a special type of function from $\mathbf{R}^n \times ... \times \mathbf{R}^n$ to the real numbers.

In order to be able to use this definition, we would have to prove that such a function on the space of matrices, satisfying conditions *a* through *c*, even exists and then that it is unique. Existence can be shown by checking that our first (inductive) definition for the determinant satisfies these conditions, though it is a painful calculation. The proof of uniqueness can be found in almost any linear algebra text.

The third definition for the determinant is the most geometric but is also the most vague. We must think of an $n \times n$ matrix A as a linear transformation from $\mathbf{R}^n$ to $\mathbf{R}^n$. Then A will map the unit cube in $\mathbf{R}^n$ to some different object (a parallelepiped). The unit cube has, by definition, a volume of one.

Definition 1.5.4 *The* determinant *of the matrix A is the signed volume of the image of the unit cube.*

This is not well-defined, as the very method of defining the volume of the image has not been described. In fact, most would define the signed volume of the image to be the number given by the determinant using one of the two earlier definitions. But this can be all made rigorous, though at the price of losing much of the geometric insight.

Let's look at some examples: the matrix $A = \begin{pmatrix} 2 & 0 \\ 0 & 1 \end{pmatrix}$ takes the unit square to

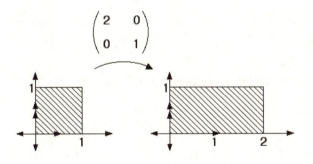

Since the area is doubled, we must have

$$\det(A) = 2.$$

Signed volume means that if the orientations of the edges of the unit cube are changed, then we must have a negative sign in front of the volume. For example, consider the matrix $A = \begin{pmatrix} -2 & 0 \\ 0 & 1 \end{pmatrix}$. Here the image is

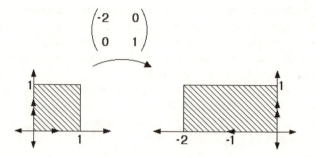

Note that the orientations of the sides are flipped. Since the area is still doubled, the definition will force

$$\det(A) = -2.$$

To rigorously define orientation is somewhat tricky (we do it in Chapter Six), but its meaning is straightforward.

The determinant has many algebraic properties. For example,

Lemma 1.5.1 : *If A and B are $n \times n$ matrices, then*

$$\det(AB) = \det(A)\det(B).$$

This can be proven by either a long calculation or by concentrating on the definition of the determinant as the change of volume of a unit cube.

1.6   The Key Theorem of Linear Algebra

Here is the the key theorem of linear algebra. (Note: we have yet to define eigenvalues and eigenvectors, but we will in section eight.)

Theorem 1.6.1 (Key Theorem) *Let A be an $n \times n$ matrix. Then the following are equivalent:*

1. *A is invertible.*

2. *$\det(A) \neq 0$.*

3. *$ker(A) = 0$.*

4. *If $\mathbf{b}$ is a column vector in $\mathbf{R}^n$, there is a unique column vector $\mathbf{x}$ in $\mathbf{R}^n$ satisfying $A\mathbf{x} = \mathbf{b}$.*

5. *The columns of A are linearly independent $n \times 1$ column vectors.*

6. *The rows of A are linearly independent $1 \times n$ row vectors.*

7. *The transpose A^t of A is invertible. (Here, if $A = (a_{ij})$, then $A^t = (a_{ji})$).*

8. *All of the eigenvalues of A are nonzero.*

We can restate this theorem in terms of linear transformations.

Theorem 1.6.2 (Key Theorem) *Let $T : V \to V$ be a linear transformation. Then the following are equivalent:*

1. *T is invertible.*

2. *$\det(T) \neq 0$, where the determinant is defined by a choice of basis on V.*

3. *$\ker(T) = 0$.*

4. *If b is a vector in V, there is a unique vector v in V satisfying $T(v) = b$.*

5. *For any basis $v_1, \ldots, v_n$ of V, the image vectors $T(v_1), \ldots, T(v_n)$ are linearly independent.*

6. *For any basis $v_1, \ldots, v_n$ of V, if S denotes the transpose linear transformation of T, then the image vectors $S(v_1), \ldots, S(v_n)$ are linearly independent.*

7. *The transpose of T is invertible. (Here the transpose is defined by a choice of basis on V).*

8. *All of the eigenvalues of T are nonzero.*

In order to make the correspondence between the two theorems clear, we must worry about the fact that we only have definitions of the determinant and the transpose for matrices, not for linear transformations. While we do not show it, both notions can be extended to linear transformations, provided a basis is chosen (in fact, provided we choose an inner product, which will be defined in Chapter Thirteen on Fourier series). But note that while the actual value $\det(T)$ will depend on a fixed basis, the condition that $\det(T) \neq 0$ does not. Similar statements hold for conditions (6) and (7). A proof is the goal of exercise 7, where you are asked to find any linear algebra book and then fill in the proof. It is unlikely that the linear algebra book will have this result as it is stated here. The act of translating is in fact part of the purpose of making this an exercise.

Each of the equivalences is important. Each can be studied on its own merits. It is remarkable that they are the same.

1.7   Similar Matrices

Recall that given a basis for an n dimensional vector space V, we can represent a linear transformation

$$T : V \to V$$

as an $n \times n$ matrix A. Unfortunately, if you choose a different basis for V, the matrix representing the linear transformation T will be quite different from the original matrix A. This section's goal is to find out a clean criterion for when two matrices actually represent the same linear transformation but under different choice of bases.

Definition 1.7.1 *Two $n \times n$ matrices A and B are* similar *if there is an invertible matrix C such that*

$$A = C^{-1}BC.$$

We want to see that two matrices are similar precisely when they represent the same linear transformation. Choose two bases for the vector space V, say $\{v_1, \ldots, v_n\}$ (the v basis) and $\{w_1, \ldots, w_n\}$ (the w basis). Let A be the matrix representing the linear transformation T for the v basis and let B be the matrix representing the linear transformation for the w basis. We want to construct the matrix C so that $A = C^{-1}BC$.

Recall that given the v basis, we can write each vector $z \in V$ as an $n \times 1$ column vector as follows: we know that there are unique scalars $a_1, \ldots, a_n$ with

$$z = a_1 v_1 + \cdots + a_n v_n.$$

We then write z, with respect to the v basis, as the column vector:

$$\begin{pmatrix} a_1 \\ \vdots \\ a_n \end{pmatrix}.$$

Similarly, there are unique scalars $b_1, \ldots, b_n$ so that

$$z = b_1 w_1 + \cdots + b_n w_n,$$

meaning that with respect to the w basis, the vector z is the column vector:

$$\begin{pmatrix} b_1 \\ \vdots \\ b_n \end{pmatrix}.$$

The desired matrix C will be the matrix such that

$$C \begin{pmatrix} a_1 \\ \vdots \\ a_n \end{pmatrix} = \begin{pmatrix} b_1 \\ \vdots \\ b_n \end{pmatrix}.$$

If $C = (c_{ij})$, then the entries c_{ij} are precisely the numbers which yield:

$$w_i = c_{i1}v_1 + \ldots + c_{in}v_n.$$

Then, for A and B to represent the same linear transformation, we need the diagram:

$$\begin{array}{ccc} \mathbf{R}^n & \xrightarrow{A} & \mathbf{R}^n \\ C \downarrow & & \downarrow C \\ \mathbf{R}^n & \xrightarrow{B} & \mathbf{R}^n \end{array}$$

to commute, meaning that $CA = BC$ or

$$A = C^{-1}BC,$$

as desired.

Determining when two matrices are similar is a type of result that shows up throughout math and physics. Regularly you must choose some coordinate system (some basis) in order to write down anything at all, but the underlying math or physics that you are interested in is independent of the initial choice. The key question becomes: what is preserved when the coordinate system is changed? Similar matrices allow us to start to understand these questions.

1.8 Eigenvalues and Eigenvectors

In the last section we saw that two matrices represent the same linear transformation, under different choices of bases, precisely when they are similar. This does not tell us, though, how to choose a basis for a vector space so that a linear transformation has a particularly decent matrix representation. For example, the diagonal matrix

$$A = \begin{pmatrix} 1 & 0 & 0 \\ 0 & 2 & 0 \\ 0 & 0 & 3 \end{pmatrix}$$

is similar to the matrix

$$B = \frac{1}{4} \begin{pmatrix} 1 & -4 & -5 \\ 1 & 8 & -1 \\ 5 & 4 & 15 \end{pmatrix},$$

but all recognize the simplicity of A as compared to B. (By the way, it is not obvious that A and B are similar; I started with A, chose a nonsingular matrix C and then used the software package Mathematica to compute $C^{-1}AC$ to get B. I did not just suddenly "see" that A and B are similar. No, I rigged it to be so.)

One of the purposes behind the following definitions for eigenvalues and eigenvectors is to give us tools for picking out good bases. There are, though, many other reasons to understand eigenvalues and eigenvectors.

Definition 1.8.1 *Let $T : V \to V$ be a linear transformation. Then a nonzero vector $v \in V$ will be an* eigenvector *of T with* eigenvalue λ, *a scalar, if*

$$T(v) = \lambda v.$$

For an $n \times n$ matrix A, a nonzero column vector $\mathbf{x} \in \mathbf{R}^n$ will be an eigenvector *with eigenvalue λ, a scalar, if*

$$A\mathbf{x} = \lambda \mathbf{x}.$$

Geometrically, a vector v is an eigenvector of the linear transformation T with eigenvalue λ if T stretches v by a factor of λ.

For example,

$$\begin{pmatrix} -2 & -2 \\ 6 & 5 \end{pmatrix} \begin{pmatrix} 1 \\ -2 \end{pmatrix} = 2 \begin{pmatrix} 1 \\ -2 \end{pmatrix},$$

and thus 2 is an eigenvalue and $\begin{pmatrix} 1 \\ -2 \end{pmatrix}$ an eigenvector for the linear transformation represented by the 2×2 matrix $\begin{pmatrix} -2 & -2 \\ 6 & 5 \end{pmatrix}$.

Luckily there is an easy way to describe the eigenvalues of a square matrix, which will allow us to see that the eigenvalues of a matrix are preserved under a similarity transformation.

Proposition 1.8.1 *A number λ will be an eigenvalue of a square matrix A if and only if λ is a root of the polynomial*

$$P(t) = \det(tI - A).$$

The polynomial $P(t) = \det(tI - A)$ is called the *characteristic polynomial* of the matrix A.

Proof: Suppose that λ is an eigenvalue of A, with eigenvector v. Then $Av = \lambda v$, or

$$\lambda v - Av = 0,$$

where the zero on the right hand side is the zero column vector. Then, putting in the identity matrix I, we have

$$0 = \lambda v - Av = (\lambda I - A)v.$$

Thus the matrix $\lambda I - A$ has a nontrivial kernel, v. By the key theorem of linear algebra, this happens precisely when

$$\det(\lambda I - A) = 0,$$

which means that λ is a root of the characteristic polynomial $P(t) = \det(tI - A)$. Since all of these directions can be reversed, we have our theorem. □

Theorem 1.8.1 *Let A and B be similar matrices. Then the characteristic polynomial of A is equal to the characteristic polynomial of B.*

Proof: For A and B to be similar, there must be an invertible matrix C with $A = C^{-1}BC$. Then

$$
\begin{aligned}
\det(tI - A) &= \det(tI - C^{-1}BC) \\
&= \det(tC^{-1}C - C^{-1}BC) \\
&= \det(C^{-1})\det(tI - B)\det(C) \\
&= \det(tI - B)
\end{aligned}
$$

using that $1 = \det(C^{-1}C) = \det(C^{-1})\det(C)$. □

Since the characteristic polynomials for similar matrices are the same, this means that the eigenvalues must be the same.

Corollary 1.8.1.1 *The eigenvalues for similar matrices are equal.*

Thus to see if two matrices are similar, one can compute to see if the eigenvalues are equal. If they are not, the matrices are not similar. Unfortunately in general, having equal eigenvalues does not force matrices to be similar. For example, the matrices

$$A = \begin{pmatrix} 1 & -7 \\ 0 & 2 \end{pmatrix}$$

and

$$B = \begin{pmatrix} 1 & 0 \\ 0 & 2 \end{pmatrix}$$

both have eigenvalues 1 and 2, but they are not similar. (This can be shown by assuming that there is an invertible two-by-two matrix C with $C^{-1}AC = B$ and then showing that $\det(C) = 0$, contradicting C's invertibility.)

Since the characteristic polynomial $P(t)$ does not change under a similarity transformation, the coefficients of $P(t)$ will also not change under a similarity transformation. But since the coefficients of $P(t)$ will themselves be (complicated) polynomials of the entries of the matrix A, we now have certain special polynomials of the entries of A that are invariant under a similarity transformation. One of these coefficients we have already seen in another guise, namely the determinant of A, as the following theorem shows. This theorem will more importantly link the eigenvalues of A to the determinant of A.

Theorem 1.8.2 *Let $\lambda_1, \ldots, \lambda_n$ be the eigenvalues, counted with multiplicity, of a matrix A. Then*

$$\det(A) = \lambda_1 \cdots \lambda_n.$$

Before proving this theorem, we need to discuss the idea of counting eigenvalues "with multiplicity". The difficulty is that a polynomial can have a root that must be counted more than once (e.g., the polynomial $(x - 2)^2$ has the single root 2 which we want to count twice). This can happen in particular to the characteristic polynomial. For example, consider the matrix

$$\begin{pmatrix} 5 & 0 & 0 \\ 0 & 5 & 0 \\ 0 & 0 & 4 \end{pmatrix}$$

which has as its characteristic polynomial the cubic

$$(t - 5)(t - 5)(t - 4).$$

For the above theorem, we would list the eigenvalues as 4, 5, and 5, hence counting the eigenvalue 5 twice.

Proof: Since the eigenvalues $\lambda_1, \ldots, \lambda_n$ are the (complex) roots of the characteristic polynomial $\det(tI - A)$, we have

$$(t - \lambda_1) \cdots (t - \lambda_n) = \det(tI - A).$$

Setting $t = 0$, we have

$$(-1)^n \lambda_1 \cdots \lambda_n = \det(-A).$$

In the matrix $(-A)$, each column of A is multiplied by (-1). Using the second definition of a determinant, we can factor out each of these (-1)s, to get

$$(-1)^n \lambda_1 \cdots \lambda_n = (-1)^n \det(A)$$

and our result. $\square$

Now finally to turn back to determining a "good" basis for representing a linear transformation. The measure of "goodness" is how close the matrix is to being a diagonal matrix. We will restrict ourselves to a special, but quite prevalent, class: symmetric matrices. By symmetric, we mean that if $A = (a_{ij})$, then we require that the entry at the ith row and jth column (a_{ij}) must equal to the entry at the jth row and the ith column (a_{ji}). Thus

$$\begin{pmatrix} 5 & 3 & 4 \\ 3 & 5 & 2 \\ 4 & 2 & 4 \end{pmatrix}$$

is symmetric but

$$\begin{pmatrix} 5 & 2 & 3 \\ 6 & 5 & 3 \\ 2 & 18 & 4 \end{pmatrix}$$

is not.

Theorem 1.8.3 *If A is a symmetric matrix, then there is a matrix B similar to A which is not only diagonal but with the entries along the diagonal being precisely the eigenvalues of A.*

Proof: The proof basically rests on showing that the eigenvectors for A form a basis in which A becomes our desired diagonal matrix. We will assume that the eigenvalues for A are distinct, as technical difficulties occur when there are eigenvalues with multiplicity.

Let $\mathbf{v}_1, \mathbf{v}_2, \ldots, \mathbf{v}_n$ be the eigenvectors for the matrix A, with corresponding eigenvalues $\lambda_1, \lambda_2, \ldots, \lambda_n$. Form the matrix

$$C = (\mathbf{v}_1, \mathbf{v}_2, \ldots, \mathbf{v}_n),$$

where the ith column of C is the column vector $\mathbf{v}_i$. We will show that the matrix $C^{-1}AC$ will satisfy our theorem. Thus we want to show that $C^{-1}AC$ equals the diagonal matrix

$$B = \begin{pmatrix} \lambda_1 & 0 & \cdots & 0 \\ \vdots & \vdots & \vdots & \vdots \\ 0 & 0 & \cdots & \lambda_n \end{pmatrix}.$$

Denote

$$\mathbf{e}_1 = \begin{pmatrix} 1 \\ 0 \\ \vdots \\ 0 \end{pmatrix}, \mathbf{e}_2 = \begin{pmatrix} 0 \\ 1 \\ \vdots \\ 0 \end{pmatrix}, \ldots, \mathbf{e}_n = \begin{pmatrix} 0 \\ 0 \\ \vdots \\ 1 \end{pmatrix}.$$

Then the above diagonal matrix B is the unique matrix with $B\mathbf{e}_i = \lambda_i \mathbf{e}_i$, for all i. Our choice for the matrix C now becomes clear as we observe that for all i, $C\mathbf{e}_i = \mathbf{v}_i$. Then we have

$$C^{-1}AC\mathbf{e}_i = C^{-1}A\mathbf{v}_i = C^{-1}(\lambda_i \mathbf{v}_i) = \lambda_i C^{-1}\mathbf{v}_i = \lambda_i \mathbf{e}_i,$$

giving us the theorem. $\square$

This is of course not the end of the story. For nonsymmetric matrices, there are other canonical ways finding "good" similar matrices, such as the Jordan canonical form, the upper triangular form and rational canonical form.

1.9   Dual Vector Spaces

It pays to study functions. In fact, functions appear at times to be more basic than their domains. In the context of linear algebra, the natural class of functions is linear transformations, or linear maps from one vector space to another. Among all real vector spaces, there is one that seems simplest, namely the one-dimensional vector space of the real numbers $\mathbf{R}$. This leads us to examine a special type of linear transformation on a vector space, those that map the vector space to the real numbers, the set of which we will call the *dual space*. Dual spaces regularly show up in mathematics.

Let V be a vector space. The *dual vector space*, or *dual space*, is:

$$
\begin{aligned}
V^* &= \{\text{linear maps from } V \text{ to the real numbers } \mathbf{R}\} \\
&= \{v^* : V \to \mathbf{R} \mid v^* \text{ is linear}\}.
\end{aligned}
$$

You can check that the dual space V^* is itself a vector space.

Let $T : V \to W$ be a linear transformation. Then we can define a natural linear transformation

$$T^* : W^* \to V^*$$

from the dual of W to the dual of V as follows. Let $w^* \in W^*$. Then given any vector w in the vector space W, we know that $w^*(w)$ will be a real number. We need to define T^* so that $T^*(w^*) \in V^*$. Thus given any vector $v \in V$, we need $T^*(w^*)(v)$ to be a real number. Simply define

$$T^*(w^*)(v) = w^*(T(v)).$$

By the way, note that the direction of the linear transformation $T : V \to W$ is indeed reversed to $T^* : W^* \to V^*$. Also by "natural", we do not mean that the map T^* is "obvious" but instead that it can be uniquely associated to the original linear transformation T.

Such a dual map shows up in many different contexts. For example, if X and Y are topological spaces with a continuous map $F : X \to Y$ and if $C(X)$ and $C(Y)$ denote the sets of continuous real-valued functions on X and Y, then here the dual map

$$F^* : C(Y) \to C(X)$$

is defined by $F^*(g)(x) = g(F(x))$, where g is a continuous map on Y.

Attempts to abstractly characterize all such dual maps were a major theme of mid-twentieth century mathematics and can be viewed as one of the beginnings of category theory.

1.10   Books

Mathematicians have been using linear algebra since they have been doing mathematics, but the styles, methods and the terminologies have shifted. For example, if you look in a college course catalogue in 1900 or probably even 1950, there will be no undergraduate course called linear algebra. Instead there were courses such as "Theory of Equations" or simply "Algebra". As seen in one of the more popular textbooks in the first part of the twentieth century, Maxime Bocher's *Introduction to Higher Algebra* [10], the concern was on concretely solving systems of linear equations. The results were written in an algorithmic style. Modern day computer programmers usually find this style of text far easier to understand than current math books. In the 1930s, a fundamental change in the way algebraic topics were taught occurred with the publication of Van der Waerden's *Modern Algebra* [113][114], which was based on lectures of Emmy Noether and Emil Artin. Here a more abstract approach was taken. The first true modern day linear algebra text was Halmos' *Finite-dimensional Vector Spaces* [52]. Here the emphasis is on the idea of a vector space from the very beginning. Today there are many beginning texts. Some start with systems of linear equations and then deal with vector spaces, others reverse the process. A long time favorite of many is Strang's *Linear Algebra and Its Applications* [109]. As a graduate student, you should volunteer to teach or TA linear algebra as soon as possible.

1.11   Exercises

1. Let $L : V \to W$ be a linear transformation between two vector spaces. Show that

$$\dim(ker(L)) + \dim(Im(L)) = \dim(V).$$

2. Consider the set of all polynomials in one variable with real coefficients of degree less than or equal to three.

    a. Show that this set forms a vector space of dimension four.

    b. Find a basis for this vector space.

    c. Show that differentiating a polynomial is a linear transformation.

    d. Given the basis chosen in part (b), write down the matrix representative of the derivative.

3. Let A and B be two $n \times n$ invertible matrices. Prove that

$$(AB)^{-1} = B^{-1}A^{-1}.$$

4. Let

$$A = \begin{pmatrix} 2 & 3 \\ 3 & 5 \end{pmatrix}$$

Find a matrix C so that $C^{-1}AC$ is a diagonal matrix.

5. Denote the vector space of all functions

$$f : \mathbf{R} \to \mathbf{R}$$

which are infinitely differentiable by $C^{\infty}(\mathbf{R})$. This space is called the space of smooth functions.

    a. Show that $C^{\infty}(\mathbf{R})$ is infinite dimensional.

    b. Show that differentiation is a linear transformation:

$$\frac{\mathrm{d}}{\mathrm{d}x} : C^{\infty}(\mathbf{R}) \to C^{\infty}(\mathbf{R}).$$

    c. For a real number λ, find an eigenvector for $\frac{\mathrm{d}}{\mathrm{d}x}$ with eigenvalue λ.

6. Let V be a finite dimensional vector space. Show that the dual vector space V^* has the same dimension as V.

7. Find a linear algebra text. Use it to prove the key theorem of linear algebra. Note that this is a long exercise but is to be taken seriously.

Chapter 2

ϵ and δ Real Analysis

Basic Object:	The Real Numbers
Basic Maps:	Continuous and Differentiable Functions
Basic Goal:	The Fundamental Theorem of Calculus

While the basic intuitions behind differentiation and integration were known by the late 1600s, allowing for a wealth of physical and mathematical applications to develop during the 1700s, it was only in the 1800s that sharp, rigorous definitions were finally given. The key concept is that of a limit, from which follow the definitions for differentiation and integration and rigorous proofs of their basic properties. Far from a mere exercise in pedantry, this rigorization actually allowed mathematicians to discover new phenomena. For example, Karl Weierstrass discovered a function that was continuous everywhere but differentiable nowhere. In other words, there is a function with no breaks but with sharp edges at every point. Key to his proof is the need for limits to be applied to sequences of functions, leading to the idea of uniform convergence.

We will define limits and then use this definition to develop the ideas of continuity, differentiation and integration of functions. Then we will show how differentiation and integration are intimately connected in the Fundamental Theorem of Calculus. Finally we will finish with uniform convergence of functions and Weierstrass' example.

2.1 Limits

Definition 2.1.1 *A function $f : \mathbf{R} \to \mathbf{R}$ has a* limit *L at the point a if given any real number $\epsilon > 0$ there is a real number $\delta > 0$ such that for all* *real numbers x with*

$$0 < |x - a| < \delta,$$

we have

$$|f(x) - L| < \epsilon.$$

This is denoted by

$$\lim_{x \to a} f(x) = L.$$

Intuitively, the function $f(x)$ should have a limit L at a point a if, for numbers x near a, the value of the function $f(x)$ is close to the number L. In other words, to guarantee that $f(x)$ be close to L, we can require that x is close to a. Thus if we want $f(x)$ to be within an arbitrary $\epsilon > 0$ of the number L (i.e., if we want $|f(x) - L| < \epsilon$), we must be able to specify how close to a we must force x to be. Therefore, given a number $\epsilon > 0$ (no matter how small), we must be able to find a number $\delta > 0$ so that if x is within δ of a, we have that $f(x)$ is within an ϵ of L. This is precisely what the definition says, in symbols.

For example, if the above definition of a limit is to make sense, it must yield that

$$\lim_{x \to 2} x^2 = 4.$$

We will check this now. It must be emphasized that we would be foolish to show that x^2 approaches 4 as x approaches 2 by actually using the definition. We are again doing the common trick of using an example whose answer we already know to check the reasonableness of a new definition. Thus for any $\epsilon > 0$, we must find a $\delta > 0$ so that if $0 < |x - 2| < \delta$, we will have

$$|x^2 - 4| < \epsilon.$$

Set

$$\delta = \min\left(\frac{\epsilon}{5}, 1\right).$$

As often happens, the initial work in finding the correct expression for δ is hidden. Also, the '5' in the denominator will be seen not to be critical. Let $0 < |x - 2| < \delta$. We want $|x^2 - 4| < \epsilon$. Now

$$|x^2 - 4| = |x - 2| \cdot |x + 2|.$$

Since x is within δ of 2,

$$|x + 2| < (2 + \delta) + 2 = 4 + \delta \le 5.$$

Thus

$$|x^2 - 4| = |x - 2| \cdot |x + 2| < 5 \cdot |x - 2| < 5 \cdot \frac{\epsilon}{5} = \epsilon.$$

We are done.

2.2 Continuity

Definition 2.2.1 *A function $f : \mathbf{R} \to \mathbf{R}$ is* continuous *at a if*

$$\lim_{x \to a} f(x) = f(a).$$

Of course, any intuition about continuous functions should capture the notion that a continuous function cannot have any breaks in its graph. In other words, you can graph a continuous function without having to lift your pencil from the page. (As with any sweeping intuition, this one will break down if pushed too hard.)

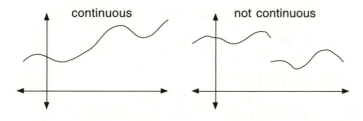

In ϵ and δ notation, the definition of continuity is:

Definition 2.2.2 *A function $f : \mathbf{R} \to \mathbf{R}$ is* continuous *at a if given any $\epsilon > 0$, there is some $\delta > 0$ such that for all x with $0 < |x - a| < \delta$, we have $|f(x) - f(a)| < \epsilon$.*

For an example, we will write down a function that is clearly not continuous at the origin 0, and use this function to check the reasonableness of the definition.
Let

$$f(x) = \begin{cases} 1 & \text{if } x > 0 \\ -1 & \text{if } x \leq 0 \end{cases}$$

Note that the graph of $f(x)$ has a break in it at the origin.

We want to capture this break by showing that

$$\lim_{x \to 0} f(x) \neq f(0).$$

Now $f(0) = -1$. Let $\epsilon = 1$ and let $\delta > 0$ be any positive number. Then for any x with $0 < x < \delta$, we have $f(x) = 1$. Then

$$|f(x) - f(0)| = |1 - (-1)| = 2 > 1 = \epsilon.$$

Thus for all positive $x < \delta$.

$$|f(x) - f(0)| > \epsilon.$$

Hence, for any $\delta > 0$, there are x with

$$|x - 0| < \delta$$

but

$$|f(x) - f(0)| > \epsilon.$$

This function is indeed not continuous.

2.3 Differentiation

Definition 2.3.1 *A function* $f : \mathbf{R} \to \mathbf{R}$ *is* differentiable *at a if*

$$\lim_{x \to a} \frac{f(x) - f(a)}{x - a}$$

exists. This limit is called the derivative *and is denoted by (among many other symbols)* $f'(a)$ *or* $\frac{\mathrm{d}f}{\mathrm{d}x}(a)$.

One of the key intuitive meanings of a derivative is that it should give the slope of the tangent line to the curve $y = f(x)$ at the point a. While logically the current definition of a tangent line must include the above definition of derivative, in pictures the tangent line is of course:

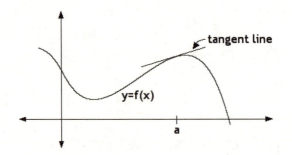

The idea behind the definition is that we can compute the slope of a line defined by any two points in the plane. In particular, for any $x \neq a$, the slope of the secant line through the points $(a, f(a))$ and $(x, f(x))$ will be

$$\frac{f(x) - f(a)}{x - a}.$$

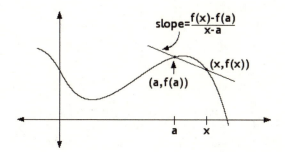

We now let x approach a. The corresponding secant lines will approach the tangent line. Thus the slopes of the secant lines must approach the slope of the tangent line.

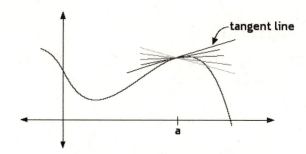

Hence the definition for the slope of the tangent line should be:

$$f'(a) = \lim_{x \to a} \frac{f(x) - f(a)}{x - a}.$$

Part of the power of derivatives (and why they can be taught to high school seniors and first year college students) is that there is a whole calculational machinery to differentiation, allowing us to usually avoid the actual taking of a limit.

We now look at an example of a function that does not have a derivative at the origin, namely

$$f(x) = |x|.$$

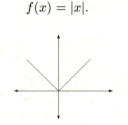

This function has a sharp point at the origin and thus no apparent tangent line there. We will show that the definition yields that $f(x) = |x|$ is indeed not differentiable at $x = 0$. Thus we want to show that

$$\lim_{x \to 0} \frac{f(x) - f(0)}{x - 0}$$

does not exist. Luckily

$$\frac{f(x) - f(0)}{x - 0} = \frac{|x|}{x} = \begin{cases} 1, & x > 0 \\ -1, & x < 0 \end{cases},$$

which we have already shown in the last section to not have a limit as x approaches 0.

2.4 Integration

Intuitively the integral of a positive function $f(x)$ with domain $a \le x \le b$ should be the area under the curve $y = f(x)$ above the x-axis.

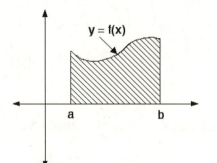

When the function $f(x)$ is not everywhere positive, then its integral should be the area under the positive part of the curve $y = f(x)$ minus the area above the negative part of $y = f(x)$.

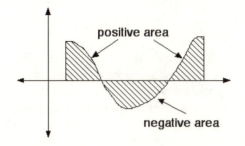

Of course this is hardly rigorous, as we do not yet even have a good definition for area.

The main idea is that the area of a rectangle with height a and width b is ab.

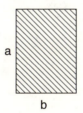

To find the area under a curve $y = f(x)$ we first find the area of various rectangles contained under the curve and then the area of various rectangles just outside the curve.

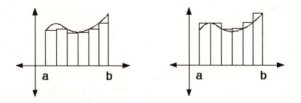

We then make the rectangles thinner and thinner, as in:

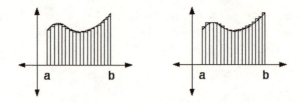

We take the limits, which should result in the area under the curve.

Now for the more technically correct definitions. We consider a real-valued function $f(x)$ with domain the closed interval $[a, b]$. We first want to divide, or partition, the interval $[a, b]$ into little segments that will be the widths of the approximating rectangles. For each positive integer n, let

$$\Delta t = \frac{b - a}{n}$$

and

$$
\begin{aligned}
a &= t_0, \\
t_1 &= t_0 + \Delta t, \\
t_2 &= t_1 + \Delta t, \\
&\vdots \\
t_n(= b) &= t_{n-1} + \Delta t.
\end{aligned}
$$

For example, on the interval $[0, 2]$ with $n = 4$, we have $\Delta t = \frac{2-0}{4} = \frac{1}{2}$ and

$$t_0 = 0 \quad t_1 = \tfrac{1}{2} \quad t_2 = 1 \quad t_3 = \tfrac{3}{2} \quad t_4 = 2$$

On each interval $[t_{k-1}, t_k]$, choose points l_k and u_k such that for all points t on $[t_{k-1}, t_k]$, we have

$$f(l_k) \leq f(t)$$

and

$$f(u_k) \geq f(t).$$

We make these choices in order to guarantee that the rectangle with base $[t_{k-1}, t_k]$ and height $f(l_k)$ is just under the curve $y = f(x)$ and that the rectangle with base $[t_{k-1}, t_k]$ and height $f(u_k)$ is just outside the curve $y = f(x)$.

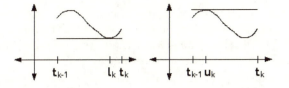

Definition 2.4.1 *Let $f(x)$ be a real-valued function defined on the closed interval $[a, b]$. For each positive integer n, let the* lower sum *of $f(x)$ be*

$$L(f, n) = \sum_{k=1}^{n} f(l_k) \Delta t$$

and the upper sum *be*

$$U(f, n) = \sum_{k=1}^{n} f(u_k) \Delta t.$$

Note that the lower sum $L(f, n)$ is the sum of the areas of the rectangles below our curve while the upper sum $U(f, n)$ is the sum of the areas of the rectangles sticking out above our curve.

Now we can define the integral.

Definition 2.4.2 *A real-valued function $f(x)$ with domain the closed interval $[a, b]$ is said to be* integrable *if the following two limits exist and are equal:*

$$\lim_{n \to \infty} L(f, n) = \lim_{n \to \infty} U(f, n).$$

If these limits are equal, we denote the limit by $\int_a^b f(x)\, dx$ and call it the integral of $f(x)$.

While from pictures it does seem that the above definition will capture the notion of an area under a curve, almost any explicit attempt to actually calculate an integral will be quite difficult. The goal of the next section, the Fundamental Theorem of Calculus, is to see how the integral (an area-finding device) is linked to the derivative (a slope-finding device). This will actually allow us to compute many integrals.

2.5 The Fundamental Theorem of Calculus

Given a real-valued function $f(x)$ defined on the closed interval $[a, b]$ we can use the above definition of integral to define a new function, via setting:

$$F(x) = \int_a^x f(t)\, dt.$$

We use the variable t inside the integral sign since the variable x is already being used as the independent variable for the function $F(x)$. Thus the value of $F(x)$ is the number that is the (signed) area under the curve $y = f(x)$ from the endpoint a to the value x.

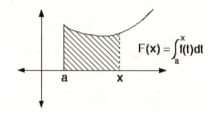

The amazing fact is that the derivative of this new function $F(x)$ will simply be the original function $f(x)$. This means that in order to find the integral of $f(x)$, you should, instead of fussing with upper and lower sums, simply try to find a function whose derivative is $f(x)$.

All of this is contained in:

Theorem 2.5.1 (Fundamental Theorem of Calculus) *Let $f(x)$ be a real-valued continuous function defined on the closed interval $[a, b]$ and define*

$$F(x) = \int_a^x f(t)\, \mathrm{d}t.$$

Then:

a) The function $F(x)$ is differentiable and

$$\frac{\mathrm{d}F(x)}{\mathrm{d}x} = \frac{\mathrm{d}\int_a^x f(t)\, \mathrm{d}t}{\mathrm{d}x} = f(x)$$

and

b) If $G(x)$ is a real-valued differentiable function defined on the closed interval $[a, b]$ whose derivative is:

$$\frac{\mathrm{d}G(x)}{\mathrm{d}x} = f(x),$$

then

$$\int_a^b f(x)\, \mathrm{d}x = G(b) - G(a).$$

First to sketch part a: We want to show that for all x in the interval $[a, b]$, the following limit exists and equals $f(x)$:

$$\lim_{h \to 0} \frac{F(x + h) - F(x)}{h} = f(x).$$

Note that we have mildly reformulated the definition of the derivative, from $\lim_{x \to x_0}(f(x) - f(x_0))/(x - x_0)$ to $\lim_{h \to 0}(f(x + h) - f(x))/h$. These are equivalent. Also, for simplicity, we will only show this for x in the open interval (a, b) and take the limit only for positive h. Consider

$$\frac{F(x + h) - F(x)}{h} = \frac{\int_a^{x+h} f(t)\, dt - \int_a^x f(t)\, dt}{h}$$

$$= \frac{\int_x^{x+h} f(t)\, dt}{h}.$$

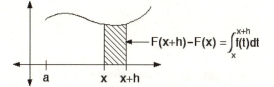

On the interval $[x, x + h]$, for each h define l_h and u_h so that for all points t on $[x, x + h]$, we have

$$f(l_h) \le f(t)$$

and

$$f(u_h) \ge f(t).$$

(Note that we are, in a somewhat hidden fashion, using that a continuous function on an interval like $[x, x + h]$ will have points such as l_h and u_h. In the chapter on point set topology, we will make this explicit, by seeing that on a compact set, such as $[x, x + h]$, a continuous function must achieve both its maximum and minimum.)

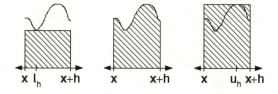

Then we have

$$f(l_h)h \le \int_x^{x+h} f(t)\, dt \le f(u_h)h.$$

Dividing by $h > 0$ gives us:

$$f(l_h) \leq \frac{\int_x^{x+h} f(t)\,dt}{h} \leq f(u_h).$$

Now both the l_h and the u_h approach the point x as h approaches zero. Since $f(x)$ is continuous, we have that

$$\lim_{h \to 0} f(l_h) = \lim_{h \to 0} f(u_h) = f(x)$$

and our result.

Turn to part b: Here we are given a function $G(x)$ whose derivative is:

$$\frac{dG(x)}{dx} = f(x).$$

Keep the notation of part a, namely that $F(x) = \int_a^x f(t)\,dt$. Note that $F(a) = 0$ and

$$\int_a^b f(t)\,dt = F(b) = F(b) - F(a).$$

By part a, we know that the derivative of $F(x)$ is the function $f(x)$. Thus the derivatives of $F(x)$ and $G(x)$ agree, meaning that

$$\frac{d(F(x) - G(x))}{dx} = f(x) - f(x) = 0.$$

But a function whose derivative is always zero must be a constant. (We have not shown this. It is quite reasonable, as the only way the slope of the tangent can always be zero is if the graph of the function is a horizontal line; the proof does take some work.) Thus there is a constant c such that

$$F(x) = G(x) + c.$$

Then

$$\int_a^b f(t)\,dt = F(b) = F(b) - F(a)$$

$$= (G(b) + c) - (G(a) + c)$$

$$= G(b) - G(a)$$

as desired.

2.6 Pointwise Convergence of Functions

Definition 2.6.1 *Let $f_n : [a, b] \to \mathbf{R}$ be a sequence of functions*

$$f_1(x), f_2(x), f_3(x), \dots$$

defined on an interval $[a, b] = \{x : a \leq x \leq b\}$. This sequence $\{f_n(x)\}$ will converge pointwise *to a function*

$$f(x) : [a, b] \to \mathbf{R}$$

if for all α in $[a, b]$,

$$\lim_{n \to \infty} f_n(\alpha) = f(\alpha).$$

In ϵ and δ notation, we would say that $\{f_n(x)\}$ *converges pointwise* to $f(x)$ if for all α in $[a, b]$ and given any $\epsilon > 0$, there is a positive integer N such that for all $n \geq N$, we have $|f(\alpha) - f_n(\alpha)| < \epsilon$.

Intuitively, a sequence of functions $f_n(x)$ will converge pointwise to a function $f(x)$ if, given any α, eventually (for huge n) the numbers $f_n(\alpha)$ become arbitrarily close to the number $f(\alpha)$. The importance of a good notion for convergence of functions stems from the frequent practice of only approximately solving a problem and then using the approximation to understand the true solution. Unfortunately, pointwise convergence is not as useful or as powerful as the next section's topic, uniform convergence, in that the pointwise limit of reasonable functions (e.g., continuous or integrable functions) does not guarantee the reasonableness of the limit, as we will see in the next example.

Here we show that the pointwise limit of continuous functions need not be continuous. For each positive integer n, set

$$f_n(x) = x^n$$

for all x on $[0, 1]$.

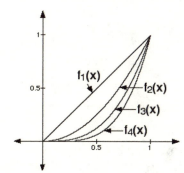

Set

$$f(x) = \begin{cases} 1, & x = 1 \\ 0, & 0 \le x < 1 \end{cases}$$

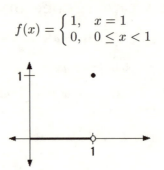

Clearly $f(x)$ is not continuous at the endpoint $x = 1$ while all of the functions $f_n(x) = x^n$ are continuous on the entire interval. But we will see that the sequence $\{f_n(x)\}$ does indeed converge pointwise to $f(x)$.

Fix α in $[0, 1]$. If $\alpha = 1$, then $f_n(1) = 1^n = 1$ for all n. Then

$$\lim_{x \to \infty} f_n(1) = \lim_{n \to \infty} 1 = 1 = f(1).$$

Now let $0 \le \alpha < 1$. We will use (without proving) the fact that for any number α less than 1, the limit of α^n will approach 0 as n approaches ∞. In particular,

$$\begin{aligned} \lim_{n \to \infty} f_n(\alpha) &= \lim_{n \to \infty} \alpha^n \\ &= 0 \\ &= f(\alpha). \end{aligned}$$

Thus the pointwise limit of a sequence of continuous functions need not be continuous.

2.7 Uniform Convergence

Definition 2.7.1 *A sequence of functions $f_n : [a, b] \to \mathbf{R}$ will converge uniformly to a function $f : [a, b] \to \mathbf{R}$ if given any $\epsilon > 0$, there is a positive integer N such that for all $n \ge N$, we have*

$$|f(x) - f_n(x)| < \epsilon$$

for all points x.

The intuition is that if we put an ϵ-tube around the function $y = f(x)$, the functions $y = f_n(x)$ will eventually fit inside this band.

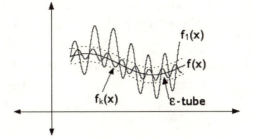

The key here is that the same ϵ and N will work for all x. This is not the case in the definition of pointwise convergence, where the choice of N depends on the number x.

Almost all of the desirable properties of the functions in the sequence will be inherited by the limit. The major exception is differentiability, but even here a partial result is true. As an example of how these arguments work, we will show

Theorem 2.7.1 *Let $f_n : [a,b] \to \mathbf{R}$ be a sequence of continuous functions converging uniformly to a function $f(x)$. Then $f(x)$ will be continuous.*

Proof: We need to show that for all α in $[a,b]$,

$$\lim_{x \to \alpha} f(x) = f(\alpha).$$

Thus, given any $\epsilon > 0$, we must find some $\delta > 0$ such that for $0 < |x - \alpha| < \delta$, we have

$$|f(x) - f(\alpha)| < \epsilon.$$

By uniform convergence, there is a positive integer N so that

$$|f(x) - f_N(x)| < \frac{\epsilon}{3}$$

for all x. (The reason for the $\frac{\epsilon}{3}$ will be seen in a moment.)

By assumption each function $f_N(x)$ is continuous at the point α. Thus there is a $\delta > 0$ such that for $0 < |x - \alpha| < \delta$, we have

$$|f_N(x) - f_N(\alpha)| < \frac{\epsilon}{3}.$$

Now to show that for $0 < |x - \alpha| < \delta$, we will have

$$|f(x) - f(\alpha)| < \epsilon.$$

We will use the trick of adding appropriate terms which sum to zero and then applying the triangle inequality ($|A + B| \leq |A| + |B|$). We have

$$
\begin{aligned}
|f(x) - f(\alpha)| &= |f(x) - f_N(x) + f_N(x) - f_N(\alpha) + f_N(\alpha) - f(\alpha)| \\
&\leq |f(x) - f_N(x)| + |f_N(x) - f_N(\alpha)| + |f_N(\alpha) - f(\alpha)| \\
&< \frac{\epsilon}{3} + \frac{\epsilon}{3} + \frac{\epsilon}{3} \\
&= \epsilon,
\end{aligned}
$$

and we are done. □

We can now make sense out of series (infinite sums) of functions.

Definition 2.7.2 *Let $f_1(x), f_2(x), \ldots$ be a sequence of functions. The series of functions*

$$
f_1(x) + f_2(x) + \ldots = \sum_{k=1}^{\infty} f_k(x)
$$

converges uniformly *to a function $f(x)$ if the sequence of partial sums: $f_1(x), f_1(x) + f_2(x), f_1(x) + f_2(x) + f_3(x), \ldots$ converges uniformly to $f(x)$.*

In terms of ϵ and $\delta's$, the infinite series of functions $\sum_{k=1}^{\infty} f_k(x)$ converges uniformly to $f(x)$ if given any $\epsilon > 0$ there is a positive integer N such that for all $n \geq N$,

$$
|f(x) - \sum_{k=1}^{n} f_k(x)| < \epsilon,
$$

for all x.

We have

Theorem 2.7.2 *If each function $f_k(x)$ is continuous and if $\sum_{k=1}^{\infty} f_k(x)$ converges uniformly to $f(x)$, then $f(x)$ must be continuous.*

This follows from the fact that the finite sum of continuous functions is continuous and the previous theorem.

The writing of a function as a series of uniformly converging (simpler) functions is a powerful method of understanding and working with functions. It is the key idea behind the development of both Taylor series and Fourier series (which is the topic of Chapter Thirteen).

2.8   The Weierstrass M-Test

If we are interested in infinite series of functions $\sum_{k=1}^{\infty} f_k(x)$, then we must be interested in knowing when the series converges uniformly. Luckily the

Weierstrass M-test provides a straightforward means for determining uniform convergence. As we will see, the key is that this theorem reduces the question of uniform convergence of $\sum_{k=1}^{\infty} f_k(x)$ to a question of when an infinite series of numbers converges, for which beginning calculus provides many tools, such as the ratio test, the root test, the comparison test, the integral test, etc.

Theorem 2.8.1 *Let $\sum_{k=1}^{\infty} f_k(x)$ be a series of functions, with each function $f_k(x)$ defined on a subset A of the real numbers. Suppose $\sum_{k=1}^{\infty} M_k$ is a series of numbers such that:*

1. *$0 \leq |f_k(x)| \leq M_k$, for all $x \in A$.*

2. *The series $\sum_{k=1}^{\infty} M_k$ converges.*

Then $\sum_{k=1}^{\infty} f_k(x)$ converges uniformly and absolutely.

By *absolute convergence*, we mean that the series of absolute values $\sum_{k=1}^{\infty} |f_k(x)|$ also converges uniformly.

Proof: To show uniform convergence, we must show that, given any $\epsilon > 0$, there exists an integer N such that for all $n \geq N$, we have

$$\left| \sum_{k=n}^{\infty} f_k(x) \right| < \epsilon,$$

for all $x \in A$. Whether or not $\sum_{k=n}^{\infty} f_k(x)$ converges, we certainly have

$$\left| \sum_{k=n}^{\infty} f_k(x) \right| \leq \sum_{k=n}^{\infty} |f_k(x)|.$$

Since $\sum_{k=1}^{\infty} M_k$ converges, we know that we can find an N so that for all $n \geq N$, we have

$$\sum_{k=n}^{\infty} M_k < \epsilon.$$

Since $0 \leq |f_k(x)| \leq M_k$, for all $x \in A$, we have

$$\left| \sum_{k=n}^{\infty} f_k(x) \right| \leq \sum_{k=n}^{\infty} |f_k(x)| \leq \sum_{k=n}^{\infty} M_k < \epsilon,$$

and we are done. $\square$

Let us look an easy example. Consider the series $\sum_{k=1}^{\infty} \frac{x^k}{k!}$, which from calculus we know to be the Taylor series for e^x. We will use the Weierstrass

M-test to show that this series converges uniformly on any interval $[-a, a]$. Here we have $f_k(x) = \frac{x^k}{k!}$. Set

$$M_k = \frac{a^k}{k!}.$$

Note that for all $x \in [-a, a]$, we have $0 < |x|^n/n! \leq a^n/n!$. Thus if we can show that the series $\sum_{k=1}^{\infty} M_k = \sum_{k=1}^{\infty} \frac{a^k}{k!}$ converges, we will have uniform convergence. By the ratio test, $\sum_{k=1}^{\infty} \frac{a^k}{k!}$ will converge if the limit of ratios

$$\lim_{k \to \infty} \frac{M_{k+1}}{M_k} = \lim_{k \to \infty} \frac{\left(\frac{a^{k+1}}{(k+1)!}\right)}{\left(\frac{a^k}{k!}\right)}$$

exists and is strictly less than one. But we have

$$\lim_{k \to \infty} \frac{\frac{a^{k+1}}{(k+1)!}}{\frac{a^k}{k!}} = \lim_{k \to \infty} \frac{a}{(k+1)} = 0.$$

Thus the Taylor series for e^x will converge uniformly on any closed interval.

2.9 Weierstrass' Example

Our goal is find a function that is continuous everywhere but differentiable nowhere. When Weierstrass first constructed such functions in the late 1800s, mathematicians were shocked and surprised. The conventional wisdom of the time was that no such function could exist. The moral of this example is that one has to be careful of geometric intuition.

We will follow closely the presentation given by Spivak in his *Calculus* [102] in Chapter 23. We need a bit of notation. Set $\{x\}$ = distance from x to the nearest integer. For example, $\{\frac{3}{4}\} = \frac{1}{4}$ and $\{1.3289\} = .3289$, etc. The graph of $\{x\}$ is:

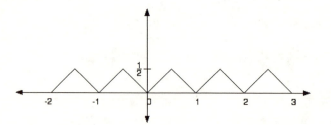

Define

$$f(x) = \sum_{k=1}^{\infty} \frac{1}{10^k} \{10^k x\}.$$

Our goal is:

Theorem 2.9.1 *The function $f(x)$ is continuous everywhere but differentiable nowhere.*

First for the intuition. For simplicity we restrict the domain to be the unit interval $(0, 1)$. For $k = 1$, we have the function $\frac{1}{10}\{10x\}$, which has a graph:

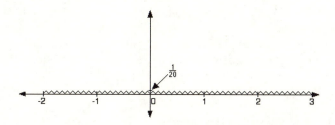

This function is continuous everywhere but not differentiable at the 19 points $.05, .1, .15, \ldots, .95$. Then $\{x\} + \frac{1}{10}\{10x\}$ has the graph:

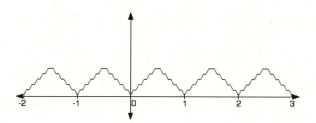

and is continuous everywhere but not differentiable at $.05, .1, .15, \ldots, .95$. For $k = 2$, the function $\frac{1}{100}\{100x\}$ is continuous everywhere but is not differentiable at its 199 sharp points. Then the partial sum $\frac{1}{10}\{10x\} + \frac{1}{100}\{100x\}$ is continuous everywhere but not differentiable at the 199 sharp points. In a similar fashion, $\frac{1}{1000}\{1000x\}$ is also continuous, but now loses differentiability at its 1999 sharp points. As we continue, at every sharp edge, we lose differentiability, but at no place is there a break in the graph. As we add all the terms in $\sum \frac{1}{10^k}\{10^k x\}$, we eventually lose differentiability at every point. The pictures are compelling, but of course we need a proof.
Proof: (We continue to follow Spivak)

The easy part is in showing that $f(x) = \sum_{k=1}^{\infty} \frac{1}{10^k} \{10^k x\}$ is continuous, as this will be a simple application of the Weierstrass M-test. We know that $\{x\} \leq \frac{1}{2}$ for all x. Thus we have, for all k, that

$$\frac{1}{10^k} \{10^k x\} \leq \frac{1}{2 \cdot 10^k}.$$

The series

$$\sum_{k=1}^{\infty} \frac{1}{2 \cdot 10^k} = \frac{1}{2} \sum_{k=1}^{\infty} \frac{1}{10^k}$$

is a geometric series and thus must converge (just use the ratio test). Then by the Weierstrass M-test, the series $f(x) = \sum_{k=1}^{\infty} \frac{1}{10^k} \{10^k x\}$ converges uniformly. Since each function $\frac{1}{10^k} \{10^k x\}$ is continuous, we have that $f(x)$ must be continuous.

It is much harder to show that $f(x)$ is not differentiable at every point; this will take some delicate work. Fix any x. We must show that

$$\lim_{h \to \infty} \frac{f(x + h) - f(x)}{h}$$

does not exist. We will find a sequence, h_m, of numbers that approach zero such that the sequence $\frac{f(x+h_m)-f(x)}{h_m}$ does not converge.

Write x in its decimal expansion:

$$x = a.a_1 a_2 \ldots,$$

where a is zero or one and each a_k is an integer between zero and nine. Set

$$h_m = \begin{cases} 10^{-m} & \text{if } a_m \neq 4 \text{ or if } a_m \neq 9 \\ -10^{-m} & \text{if } a_m = 4 \text{ or if } a_m = 9 \end{cases}$$

Then

$$x + h_m = \begin{cases} a.a_1 \ldots (a_m + 1)a_{m+1} \ldots & \text{if } a_m \neq 4 \text{ or if } a_m \neq 9 \\ a.a_1 \ldots (a_m - 1)a_{m+1} \ldots & \text{if } a_m = 4 \text{ or if } a_m = 9 \end{cases}$$

We will be looking at various $10^n(x + h_m)$. The 10^n factor just shifts where the decimal point lands. In particular, if $n > m$, then

$$10^n(x + h_m) = aa_1 \ldots (a_m \pm 1)a_{m+1} \ldots a_n.a_{n+1} \ldots,$$

in which case

$$\{10^n(x + h_m)\} = \{10^n x\}.$$

If $n \leq m$, then $10^n(x + h_m) = aa_1 \ldots a_n.a_{n+1} \ldots (a_m \pm 1)a_{m+1} \ldots$, in which case we have

$$\{10^n(x + h_m)\} = \begin{cases} 0.a_{n+1} \ldots (a_m + 1)a_{m+1} \ldots & \text{if } a_m \neq 4 \text{ or if } a_m \neq 9 \\ 0.a_{n+1} \ldots (a_m - 1)a_{m+1} \ldots & \text{if } a_m = 4 \text{ or if } a_m = 9 \end{cases}$$

We are interested in the limit of

$$\frac{f(x+h_m) - f(x)}{h_m} = \sum_{k=0}^{\infty} \frac{\frac{1}{10^k}\{10^k(x+h_m)\} - \frac{1}{10^k}\{10^k x\}}{h_m}.$$

Since $\{10^k(x+h_m)\} = \{10^k x\}$, for $k > m$, the above infinite series is actually the finite sum:

$$\sum_{k=0}^{m} \frac{\frac{1}{10^k}\{10^k(x+h_m)\} - \frac{1}{10^k}\{10^k x\}}{h_m} = \sum_{k=0}^{m} \pm 10^{m-k}(\{10^k(x+h_m)\} - \{10^k x\}).$$

We will show that each $\pm 10^{m-k}(\{10^k(x+h_m)\} - \{10^k x\})$ is a plus or minus one. Then the above finite sum is a sum of plus and minus ones and thus cannot be converging to a number, showing that the function is not differentiable.

There are two cases. Still following Spivak, we will only consider the case when $10^k x = .a_{k+1}\ldots < \frac{1}{2}$ (the case when $.a_{k+1}\ldots \geq \frac{1}{2}$ is left to the reader). Here is why we had to break our definition of the h_m into two separate cases. By our choice of h_m, $\{10^k(x+h_m)\}$ and $\{10^k x\}$ differ only in the $(m-k)$th term of the decimal expansion. Thus

$$\{10^k(x+h_m)\} - \{10^k x\} = \pm\frac{1}{10^{m-k}}.$$

Then $10^{m-k}(\{10^k(x+h_m)\} - \{10^k x\})$ will be, as predicted, a plus or minus one. $\square$

2.10   Books

The development of ϵ and δ analysis was one of the main triumphs of 1800s mathematics; this means that undergraduates for most of the last hundred years have had to learn these techniques. There are many texts. The one that I learned from and one of my favorite math books of all times is Michael Spivak's *Calculus* [102]. Though called a calculus book, even Spivak admits, in the preface to the second and third editions, that a more apt title would be "An Introduction to Real Analysis". The exposition is wonderful and the problems are excellent.

Other texts for this level of real analysis include books by Bartle [6], Berberian [7], Bressoud [13], Lang [80], Protter and Morrey [94] and Rudin [96], among many others.

2.11 Exercises

1. Let $f(x)$ and $g(x)$ be differentiable functions. Using the definition of derivatives, show

    a. $(f+g)' = f' + g'$.

    b. $(fg)' = f'g + fg'$.

    c. Assume that $f(x) = c$, where c is a constant. Show that the derivative of $f(x)$ is zero.

2. Let $f(x)$ and $g(x)$ be integrable functions.

    a. Using the definition of integration, show that the sum $f(x) + g(x)$ is an integrable function.

    b. Using the Fundamental Theorem of Calculus and problem 1.a, show that the sum $f(x) + g(x)$ is an integrable function.

3. The goal of this problem is to calculate $\int_0^1 x \, dx$ three ways. The first two methods are not supposed to be challenging.

    a. Look at the graph of the function $y = x$. Note what type of geometric object this is, and then get the area under the curve.

    b. Find a function $f(x)$ such that $f'(x) = x$ and then use the Fundamental Theorem of Calculus to find $\int_0^1 x \, dx$.

    c. This has two parts. First show by induction that

$$\sum_{i=1}^{n} i = \frac{n(n+1)}{2}.$$

Then use the definition of the integral to find $\int_0^1 x \, dx$.

4. Let $f(x)$ be differentiable. Show that $f(x)$ must be continuous. (Note: intuitively this makes a lot of sense; after all, if the function f has breaks in its graph, it should not then have well-defined tangents. This problem is an exercise in the definitions.)

5. On the interval $[0, 1]$, define

$$f(x) = \begin{cases} 1 & \text{if } x \text{ is rational} \\ 0 & \text{if } x \text{ is not rational} \end{cases}$$

Show that $f(x)$ is not integrable. (Note: you will need to use the fact that any interval of any positive length must contain a rational number and an irrational number. In other words, both the rational and the irrational numbers are dense.)

6. This is a time-consuming problem but is very worthwhile. Find a calculus textbook. Go through its proof of the chain-rule, namely that

$$\frac{d}{dx} f(g(x)) = f'(g(x)) \cdot g'(x).$$

7. Go again to the calculus book that you used in problem six. Find the chapter on infinite series. Go carefully through the proofs for the following tests for convergence: the integral test, the comparison test, the limit comparison test, the ratio test and the root test. Put all of these tests into the language of ϵ and δ real analysis.

Chapter 3

Calculus for Vector-Valued Functions

Basic Object:	$\mathbf{R}^n$
Basic Map:	Differentiable functions $f : \mathbf{R}^n \to \mathbf{R}^m$
Basic Goal:	Inverse Function Theorem

3.1 Vector-Valued Functions

A function $f : \mathbf{R}^n \to \mathbf{R}^m$ is called *vector-valued* since for any vector x in $\mathbf{R}^n$, the value (or image) of $f(x)$ is a vector in $\mathbf{R}^m$. If $(x_1, \ldots, x_n)$ is a coordinate system for $\mathbf{R}^n$, the function f can be described in terms of m real-valued functions by simply writing:

$$f(x_1, \ldots, x_n) = \begin{pmatrix} f_1(x_1, \cdots, x_n) \\ \vdots \\ f_m(x_1 \ldots, x_n) \end{pmatrix}$$

Such functions occur everywhere. For example, let $f : \mathbf{R} \to \mathbf{R}^2$ be defined as

$$f(t) = \begin{pmatrix} \cos(t) \\ \sin(t) \end{pmatrix}.$$

Here t is the coordinate for $\mathbf{R}$. Of course this is just the unit circle parametrized by its angle with the x-axis.

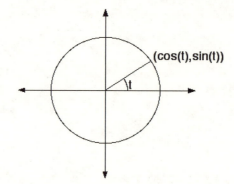

This can also be written as $x = \cos(t)$ and $y = \sin(t)$.

For another example, consider the function $f : \mathbf{R}^2 \to \mathbf{R}^3$ given by

$$f(x_1, x_2) = \begin{pmatrix} \cos x_1 \\ \sin x_1 \\ x_2 \end{pmatrix}.$$

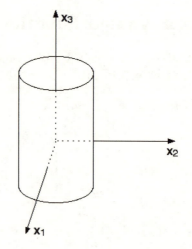

This function f maps the (x_1, x_2) plane to a cylinder in space.

Most examples are quite a bit more complicated, too complicated for pictures to even be drawn, much less used.

3.2 Limits and Continuity of Vector-Valued Functions

The key idea in defining limits for vector-valued functions is that the Pythagorean Theorem gives a natural way for measuring distance in $\mathbf{R}^n$.

Definition 3.2.1 *Let* $a = (a_1, \ldots, a_n)$ *and* $b = (b_1, \ldots, b_n)$ *be two points in* $\mathbf{R}^n$. *Then the* distance *between a and b, denoted by $|a - b|$, is*

$$|a - b| = \sqrt{(a_1 - b_1)^2 + (a_2 - b_2)^2 + \cdots + (a_n - b_n)^2}.$$

The length *of a is defined by*

$$|a| = \sqrt{a_1^2 + \cdots + a_n^2}.$$

Note that we are using the word "length" since we can think of the point a in $\mathbf{R}^n$ as a vector from the origin to the point.

Once we have a notion of distance, we can apply the standard tools from ϵ and δ style real analysis. For example, the reasonable definition of limit must be:

Definition 3.2.2 *The function* $f : \mathbf{R}^n \to \mathbf{R}^m$ *has* limit

$$L = (L_1, \ldots, L_m) \in \mathbf{R}^m$$

at the point $a = (a_1, \ldots, a_n) \in \mathbf{R}^n$ *if given any* $\epsilon > 0$, *there is some* $\delta > 0$ *such that for all* $x \in \mathbf{R}^n$, *if*

$$0 < \mid x - a \mid < \delta,$$

we have

$$|f(x) - L| < \epsilon.$$

We denote this limit by

$$\lim_{x \to a} f(x) = L$$

or by $f(x) \to L$ *as* $x \to a$.

Of course, continuity must now be defined by:

Definition 3.2.3 *The function* $f : \mathbf{R}^n \to \mathbf{R}^m$ *is* continuous *at a point a in* $\mathbf{R}^n$ *if* $\lim_{x \to a} f(x) = f(a)$.

Both the definitions of limit and continuity rely on the existence of a distance. Given different norms (distances) we will have corresponding definitions for limits and for continuity.

3.3   Differentiation and Jacobians

For single variable functions, the derivative is the slope of the tangent line (which is, recall, the best linear approximation to the graph of the original function) and can be used to find the equation for this tangent line. In a similar fashion, we want the derivative of a vector-valued function to be a tool that can be used to find the best linear approximation to the function.

We will first give the definition for the vector-valued derivative and then discuss the intuitions behind it. In particular we want this definition for vector-valued functions to agree with the earlier definition of a derivative for the case of single variable real-valued functions.

Definition 3.3.1 *A function $f : \mathbf{R}^n \to \mathbf{R}^m$ is differentiable at $a \in \mathbf{R}^n$ if there is an $m \times n$ matrix $A : \mathbf{R}^n \to \mathbf{R}^m$ such that*

$$\lim_{x \to a} \frac{|f(x) - f(a) - A \cdot (x - a)|}{|x - a|} = 0.$$

If such a limit exists, the matrix A is denoted by $Df(a)$ and is called the Jacobian

Note that $f(x), f(a)$ and $A \cdot (x - a)$ are all in $\mathbf{R}^m$ and hence

$$|f(x) - f(a) - A \cdot (x - a)|$$

is the length of a vector in $\mathbf{R}^m$. Likewise, $x - a$ is a vector in $\mathbf{R}^n$, forcing $|x - a|$ to be the length of a vector in $\mathbf{R}^n$. Further, usually there is an easy way to compute the matrix A, which we will see in a moment. Also, if the Jacobian matrix $Df(a)$ exists, one can show that it is unique, up to change of bases for $\mathbf{R}^n$ and $\mathbf{R}^m$.

We definitely want this definition to agree with the usual definition of derivative for a function $f : \mathbf{R} \to \mathbf{R}$. With $f : \mathbf{R} \to \mathbf{R}$, recall that the derivative $f'(a)$ was defined to be the limit

$$f'(a) = \lim_{x \to a} \frac{f(x) - f(a)}{x - a}.$$

Unfortunately, for a vector-valued function $f : \mathbf{R}^n \to \mathbf{R}^m$ with n and m larger than one, this one-variable definition is nonsensical, since we cannot divide vectors. We can, however, algebraically manipulate the above one-variable limit until we have a statement that can be naturally generalized to functions $f : \mathbf{R}^n \to \mathbf{R}^m$ and which will agree with our definition.

Return to the one-variable case $f : \mathbf{R} \to \mathbf{R}$. Then

$$f'(a) = \lim_{x \to a} \frac{f(x) - f(a)}{x - a}$$

is true if and only if

$$0 = \lim_{x \to a} \frac{f(x) - f(a)}{x - a} - f'(a),$$

which is equivalent to

$$0 = \lim_{x \to a} \frac{f(x) - f(a) - f'(a)(x - a)}{x - a}$$

or

$$0 = \lim_{x \to a} \frac{|f(x) - f(a) - f'(a)(x - a)|}{|x - a|}.$$

This last statement, at least formally, makes sense for functions $f : \mathbf{R}^n \to \mathbf{R}^m$, provided we replace $f'(a)$ (a number and hence a 1×1 matrix) by an $m \times n$ matrix, namely the Jacobian $Df(a)$.

As with the one-variable derivative, there is a (usually) straightforward method for computing the derivative without resorting to the actual taking of a limit, allowing us to actually calculate the Jacobian.

Theorem 3.3.1 *Let the function $f : \mathbf{R}^n \to \mathbf{R}^m$ be given by the m differentiable functions $f_1(x_1, \ldots, x_n), \ldots, f_m(x_1, \ldots, x_n)$, so that*

$$f(x_1, \ldots, x_n) = \begin{pmatrix} f_1(x_1, \ldots, x_n) \\ \vdots \\ f_m(x_1, \ldots, x_n) \end{pmatrix}$$

Then f is differentiable and the Jacobian is

$$Df(x) = \begin{pmatrix} \frac{\partial f_1}{\partial x_1} & \cdots & \frac{\partial f_1}{\partial x_n} \\ \vdots & & \vdots \\ \frac{\partial f_m}{\partial x_1} & \cdots & \frac{\partial f_m}{\partial x_n} \end{pmatrix}$$

The proof, found in most books on vector calculus, is a relatively straightforward calculation stemming from the definition of partial derivatives. But to understand it, we look at the following example. Consider our earlier example of the function $f : \mathbf{R}^2 \to \mathbf{R}^3$ given by

$$f(x_1, x_2) = \begin{pmatrix} \cos x_1 \\ \sin x_1 \\ x_2 \end{pmatrix},$$

which maps the (x_1, x_2) plane to a cylinder in space. Then the Jacobian, the derivative of this vector-valued function, will be

$$Df(x_1, x_2) \;=\; \begin{pmatrix} \partial \cos(x_1)/\partial x_1 & \partial \cos(x_1)/\partial x_2 \\ \partial (\sin x_1)/\partial x_1 & \partial \sin(x_1)/\partial x_2 \\ \partial x_2/\partial x_1 & \partial x_2/\partial x_2 \end{pmatrix}$$

$$=\; \begin{pmatrix} -sin x_1 & 0 \\ cos x_1 & 0 \\ 0 & 1 \end{pmatrix}.$$

One of the most difficult concepts and techniques in beginning calculus is the chain rule, which tells us how to differentiate the composition of two functions. For vector-valued forms, the chain rule can be easily stated (though we will not give the proof here). It should relate the derivative of the composition of functions with the derivatives of each component part and in fact has a quite clean flavor, namely:

Theorem 3.3.2 *Let* $f : \mathbf{R}^n \to \mathbf{R}^m$ *and* $g : \mathbf{R}^m \to \mathbf{R}^l$ *be differentiable functions. Then the composition function*

$$g \circ f : \mathbf{R}^n \to \mathbf{R}^l$$

is also differentiable with derivative given by: if $f(a) = b$, *then*

$$D(g \circ f)(a) = D(g)(b) \cdot D(f)(a).$$

Thus the chain rule says that to find the derivative of the composition $g \circ f$, one multiplies the Jacobian matrix for g times the Jacobian matrix for f.

One of the key intuitions behind the one-variable derivative is that $f'(a)$ is the slope of the tangent line to the curve $y = f(x)$ at the point $(a, f(a))$ in the plane $\mathbf{R}^2$. In fact, the tangent line through $(a, f(a))$ will have the equation

$$y = f(a) + f'(a)(x - a).$$

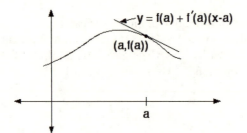

This line $y = f(a) + f'(a)(x - a)$ is the closest linear approximation to the function $y = f(x)$ at $x = a$.

Thus a reasonable criterion for the derivative of $f : \mathbf{R}^n \to \mathbf{R}^m$ should be that we can use this derivative to find a linear approximation to the geometric object $y = f(x)$, which lies in the space $\mathbf{R}^{n+m}$. But this is precisely what the definition

$$\lim_{x \to a} \frac{\mid f(x) - f(a) - Df(a)(x - a) \mid}{\mid x - a \mid} = 0$$

does. Namely, $f(x)$ is approximately equal to the linear function

$$f(a) + Df(a) \cdot (x - a).$$

Here $Df(a)$, as an $m \times n$ matrix, is a linear map from $\mathbf{R}^n \to \mathbf{R}^m$ and $f(a)$, as an element of $\mathbf{R}^m$, is a translation. Thus the vector $y = f(x)$ can be approximated by

$$y \approx f(a) + Df(a) \cdot (x - a).$$

3.4 The Inverse Function Theorem

Matrices are easy to understand, while vector-valued functions can be quite confusing. As seen in the last section, one of the points of having a derivative for vector-valued functions is that we can approximate the original function by a matrix, namely the Jacobian. The general question is now how good of an approximation do we have. What decent properties for matrices can be used to get corresponding decent properties for vector-valued functions?

This type of question could lead us to the heart of numerical analysis. We will limit ourselves to seeing that if the derivative matrix (the Jacobian) is invertible, then the original vector-valued function must also have an inverse, at least locally. This theorem, and its close relative the Implicit Function Theorem, are key technical tools that appear throughout mathematics.

Theorem 3.4.1 (Inverse Function Theorem) *For a vector-valued continuously differentiable function $f : \mathbf{R}^n \to \mathbf{R}^m$, assume that $\det Df(a) \neq 0$, at some point a in $\mathbf{R}^n$. Then there is an open neighborhood U of a in $\mathbf{R}^n$ and an open neighborhood V of $f(a)$ in $\mathbf{R}^m$ such that $f : U \to V$ is one to one, onto and has a differentiable inverse $g : V \to U$ (i.e., $g \circ f : U \to U$ is the identity and $f \circ g : V \to V$ is the identity).*

Why should a function f have an inverse? Let us think of f as being approximated by the linear function

$$f(x) \approx f(a) + Df(a) \cdot (x - a).$$

From the key theorem of linear algebra, the matrix $Df(a)$ is invertible if and only if $\det Df(a) \neq 0$. Thus $f(x)$ should be invertible if $f(a) + Df(a) \cdot (x-a)$ is invertible, which should happen precisely when $\det Df(a) \neq 0$. In fact, consider

$$y = f(a) + Df(a) \cdot (x - a).$$

Here the vector y is written explicitly as a function of the variable vector x. But if the inverse to $Df(a)$ exists, then we can write x explicitly as a function of y, namely as:

$$x = a + Df(a)^{-1} \cdot (y - f(a)).$$

In particular, we should have, if the inverse function is denoted by f^{-1}, that its derivative is simply the inverse of the derivative of the original function f, namely

$$Df^{-1}(b) = Df(a)^{-1},$$

where $b = f(a)$. This follows from the chain rule and since the composition is $f^{-1} \circ f = I$.

For the case of $f : \mathbf{R} \to \mathbf{R}$, the idea behind the Inverse Function Theorem can be captured in pictures:

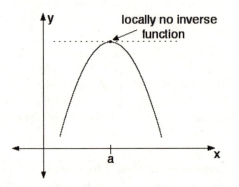

If the slope of the tangent line, $f'(a)$, is not zero, the tangent line will not be horizontal, and hence there will be an inverse.

In the statement of the theorem, we used the technical term "open set". There will be much more about this in the next chapter on topology. For now, think of an open set as a technical means allowing us to talk about all points near the points a and $f(a)$. More precisely, by an open neighborhood U of a point a in $\mathbf{R}^n$, we mean that, given any $a \in U$, there is a (small) positive ϵ such that

$$\{x : |x - a| < \epsilon\} \subset U.$$

In pictures, for example,

$$\{(x,y) \in \mathbf{R}^2 : |(x,y) - (0,0)| = \sqrt{x^2 + y^2} \leq 1\}$$

is not open (it is in fact closed, meaning that its complement is open in the plane $\mathbf{R}^2$),

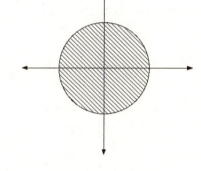

while the set

$$\{(x,y) \in \mathbf{R}^2 : |(x,y) - (0,0)| < 1\}$$

is open.

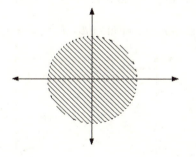

10-29-96

3.5   Implicit Function Theorem

Rarely can a curve in the plane be described as the graph of a one-variable function

$$y = f(x),$$

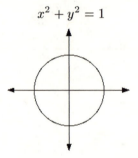

though much of our early mathematical experiences are with such functions. For example, it is impossible to write the circle

$$x^2 + y^2 = 1$$

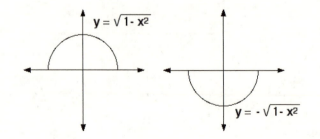

as the graph of a one-variable function, since for any value of x (besides -1 and 1) there are either no corresponding values of y on the circle or two corresponding values of y on the circle. This is unfortunate. Curves in the plane that can be cleanly written as $y = f(x)$ are simply easier to work with.

However, we can split the circle into its top and bottom halves.

For each half, the variable y can be written as a function of x: for the top half, we have

$$y = \sqrt{1 - x^2},$$

and for the bottom half,

$$y = -\sqrt{1 - x^2}.$$

Only at the two points $(1, 0)$ and $(-1, 0)$ are there problems. The difficulty can be traced to the fact that at these two points (and only at these two points) the tangent lines of the circle are perpendicular to the x-axis.

This is the key. The tangent line of a circle is the best linear approximation to the circle. If the tangent line can be written as

$$y = mx + b,$$

then it should be no surprise that the circle can be written as $y = f(x)$, at least locally.

The goal of the Implicit Function Theorem is to find a computational tool that will allow us to determine when the zero locus of a bunch of functions in some $\mathbf{R}^N$ can locally be written as the graph of a function and thus in the form $y = f(x)$, where the x denote the independent variables and the y will denote the dependent variables. Buried (not too deeply) is the intuition that we want to know about the tangent space of the zero locus of functions.

The notation is a bit cumbersome. Label a coordinate system for $\mathbf{R}^{n+k}$ by

$$x_1, \ldots, x_n, y_1, \ldots, y_k$$

which we will frequently abbreviate as (x, y). Let

$$f_1(x_1, \ldots, x_n, y_1, \ldots, y_k), \ldots, f_k(x_1, \ldots, x_n, y_1, \ldots, y_k)$$

be k continuously differentiable functions, which will frequently be written as

$$f_1(x, y), \ldots, f_k(x, y).$$

Set

$$V = \{(x, y) \in \mathbf{R}^{n+m} : f_1(x, y) = 0, \ldots, f_k(x, y) = 0\}.$$

We want to determine when, given a point $(a, b) \in V$ (where $a \in \mathbf{R}^n$ and $b \in \mathbf{R}^k$), there are k functions

$$\rho_1(x_1, \ldots, x_n), \ldots, \rho_k(x_1, \ldots, x_n)$$

defined in a neighborhood of the point a on $\mathbf{R}^n$ such that V can be described, in a neighborhood of (a, b) on $\mathbf{R}^{n+k}$, as

$$\{(x, y) \in \mathbf{R}^{n+k} : y_1 = \rho_1(x_1, \ldots, x_n), \ldots, y_k = \rho_k(x_1, \ldots, x_n)\},$$

which of course is frequently written in the shorthand of

$$V = \{y_1 = \rho_1(x), \ldots, y_k = \rho_k(x)\},$$

or even more succinctly as

$$V = \{y = \rho(x)\}.$$

Thus we want to find k functions $\rho_1, \ldots, \rho_k$ such that for all $x \in \mathbf{R}^n$, we have

$$f_1(x, \rho_1(x)) = 0, \ldots, f_k(x, \rho_k(x)) = 0.$$

Thus we want to know when the k functions $f_1, \ldots, f_k$ can be used to define (implicitly, since it does take work to actually construct them) the k functions $\rho_1, \ldots, \rho_k$.

Theorem 3.5.1 (Implicit Function Theorem) *Let $f_1(x, y), \ldots, f_k(x, y)$ be k continuously differentiable functions on $\mathbf{R}^{n+k}$ and suppose that $p = (a, b) \in \mathbf{R}^{n+k}$ is a point for which*

$$f_1(a, b) = 0, \ldots, f_k(a, b) = 0.$$

Suppose that at the point p the $k \times k$ matrix

$$M = \begin{pmatrix} \frac{\partial f_1}{\partial y_1(p)} & \cdots & \frac{\partial f_1}{\partial y_k(p)} \\ \vdots & & \vdots \\ \frac{\partial f_k}{\partial y_1(p)} & \cdots & \frac{\partial f_k}{\partial y_k(p)} \end{pmatrix}$$

is invertible. Then in a neighborhood of a in $\mathbf{R}^n$ there are k unique, differentiable functions

$$\rho_1(x), \ldots, \rho_k(x)$$

such that

$$f_1(x, \rho_1(x)) = 0, \ldots, f_k(x, \rho_k(x)) = 0.$$

Return to the circle. Here the function is $f(x, y) = x^2 + y^2 - 1 = 0$. The matrix M in the theorem will be the 1×1 matrix:

$$\frac{\partial f}{\partial y_1} = 2y.$$

This matrix is not invertible (the number is zero) only where $y = 0$, namely at the two points $(1, 0)$ and $(-1, 0)$: only at these two points will there not be an implicitly defined function ρ.

Now to sketch the main ideas of the proof, whose outline we got from [103] In fact, this theorem is a fairly easy consequence of the Inverse Function Theorem. For ease of notation, write the k-tuple $(f_1(x, y), \ldots, f_k(x, y))$ as $f(x, y)$. Define a new function $F : \mathbf{R}^{n+k} \to \mathbf{R}^{n+k}$ by

$$F(x, y) = (x, f(x, y)).$$

The Jacobian of this map is the $(n + k) \times (n + k)$ matrix

$$\begin{pmatrix} I & 0 \\ * & M \end{pmatrix}.$$

Here the I is the $n \times n$ identity matrix, M is the $k \times k$ matrix of partials as in the theorem, 0 is the $n \times k$ zero matrix and $*$ is some $k \times n$ matrix. Then the determinant of the Jacobian will be the determinant of the matrix M; hence the Jacobian is invertible if and only if the matrix M is invertible. By the Inverse Function Theorem, there will be a map $G : \mathbf{R}^{n+k} \to \mathbf{R}^{n+k}$ which will locally, in a neighborhood of the point (a, b), be the inverse of the map $F(x, y) = (x, f(x, y))$.

Let this inverse map $G : \mathbf{R}^{n+k} \to \mathbf{R}^{n+k}$ be described by the real-valued functions $G_1, \ldots, G_{n+k}$ and thus as

$$G(x, y) = (G_1(x, y), \ldots, G_{n+k}(x, y)).$$

By the nature of the map F, we see that for $1 \le i \le n$,

$$G_i(x, y) = x_i.$$

Relabel the last k functions that make up the map G by setting

$$\rho_i(x, y) = G_{i+n}(x, y).$$

Thus

$$G(x, y) = (x_1, \ldots, x_n, \rho_1(x, y), \ldots, \rho_k(x, y)).$$

We want to show that the functions $\rho_i(x, 0)$ are the functions the theorem requires.

We have yet looked at the set of points in $\mathbf{R}^{n+k}$ where the original k functions f_i are zero, namely the set that we earlier called V. The image of V under the map F will be contained in the set $(x, 0)$. Then the image $G(x, 0)$, at least locally around (a, b), will be V. Thus we must have

$$f_1(G(x, 0)) = 0, \ldots, f_k(G(x, 0)) = 0.$$

But this just means that

$$f_1(x, \rho_1(x, 0)) = 0, \ldots, f_k(x, \rho_k(x, 0)) = 0,$$

which is exactly what we wanted to show.

Here we used the Inverse Function Theorem to prove the Implicit Function Theorem. It is certainly possible and no harder to prove the Implicit Function Theorem first and then use it to prove the Inverse Function Theorem.

3.6   Books

An excellent recent book on vector calculus (and for linear algebra and Stokes' Theorem) is by Hubbard and Hubbard [64]. Fleming [37] has been the standard reference for many years. Another, more abstract approach, is in Spivak's *Calculus on Manifolds* [103]. Information on vector calculus for three variable functions is in most calculus books. A good general exercise is to look in a calculus text and translate the given results into the language of this section.

3.7   Exercises

1. In the plane $\mathbf{R}^2$ there are two natural coordinate systems: polar coordinates (r, θ) with r the radius and θ the angle with the x-axis and Cartesian coordinates (x, y).

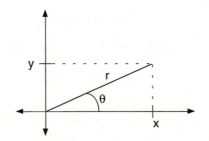

The functions that give the change of variables from polar to Cartesian coordinates are:

$$x = f(r, \theta) = r\cos(\theta)$$

$$y = g(r, \theta) = r\sin(\theta).$$

a. Compute the Jacobian of this change of coordinates.

b. At what points is the change of coordinates not well-defined (i.e., at what points is the change of coordinates not invertible)?

c. Give a geometric justification for your answer in part b.

2. There are two different ways of describing degree two monic polynomials in one variable: either by specifying the two roots or by specifying the coefficients. For example, we can describe the same polynomial by either stating that the roots are 1 and 2 or by writing it as $x^2 - 3x + 2$. The relation between the roots r_1 and r_2 and the coefficients a and b can be determined by noting that

$$(x - r_1)(x - r_2) = x^2 + ax + b.$$

Thus the space of all monic, degree two polynomials in one variable can be described by coordinates in the root space (r_1, r_2) or by coordinates in the coefficient space (a, b).

a. Write down the functions giving the change of coordinates from the root space to the coefficient space.

b. Compute the Jacobian of the coordinate change.

c. Find where this coordinate change is not invertible.

d. Give a geometric interpretation to your answer in part c.

3. Using the notation in the second question:

a. Via the quadratic equation, write down the functions giving the change of coordinates from the coordinate space to the root space.

b-d. Answer the same questions as in problem 2, but now for this new coordinate change.

4. Set $f(x, y) = x^2 - y^2$.

a. Graph the curve $f(x, y) = 0$.

b. Find the Jacobian of the function $f(x, y)$ at the point $(1, 1)$. Give a geometric interpretation of the Jacobian at this point.

c. Find the Jacobian of the function $f(x, y)$ at the point $(0, 0)$. Give a geometric interpretation for why the Jacobian is here the two-by-two zero matrix.

5. Set $f(x, y) = x^3 - y^2$.

a. Graph the curve $f(x, y) = 0$.

b. Find the Jacobian of the function $f(x, y)$ at the point $(1, 1)$. Give a geometric interpretation of the Jacobian at this point.

c. Find the Jacobian of the function $f(x, y)$ at the point $(0, 0)$. Give a geometric interpretation for why the Jacobian is here the two-by-two zero matrix.

Chapter 4

Point Set Topology

Basic Object:	Topological spaces
Basic Map:	Continuous functions

Historically, much of point set topology was developed to understand the correct definitions for such notions as continuity and dimension. By now, though, these definitions permeate mathematics, frequently in areas seemingly far removed from the traditional topological space $\mathbf{R}^n$. Unfortunately, it is not at first apparent that these more abstract definitions are at all useful; there needs to be an initial investment in learning the basic terms. In the first section, these basic definitions are given. In the next section, these definitions are applied to the topological space $\mathbf{R}^n$, where all is much more down to earth. Then we look at metric spaces. The last section applies these definitions to the Zariski topology of a commutative ring, which, while natural in algebraic geometry and algebraic number theory, is not at all similar to the topology of $\mathbf{R}^n$.

4.1 Basic Definitions

Much of point set topology consists in developing a convenient language to talk about when various points in a space are near to one another and about the notion of continuity. The key is that the same definitions can be applied to many disparate branches of math.

Definition 4.1.1 *Let X be a set of points. A collection of subsets* $\mathbf{U} = \{U_\alpha\}$ *forms a* topology *on X if*

1. Any arbitrary union of the U_α is another set in the collection $\mathbf{U}$.

2. *The intersection of any finite number of sets U_α in the collection $\mathbf{U}$ is another set in $\mathbf{U}$.*

3. *Both the empty set ϕ and the whole space X must be in $\mathbf{U}$.*

The $(X, \mathbf{U})$ is called a topological space.

The sets U_α in the collection $\mathbf{U}$ are called *open sets*. A set C is *closed* if its complement $X - C$ is open.

Definition 4.1.2 *Let A be a subset of a topological space X. Then the induced topology on A is described by letting the open sets on A be all sets of the form $U \cap A$, where U is an open set in X.*

A collection $\Sigma = \{U_\alpha\}$ of open sets is said to be an *open cover* of a subset A if A is contained in the union of the U_α.

Definition 4.1.3 *The subset A of a topological space X is* compact *if given any open cover of A, there is a finite subcover.*

In other words, if $\Sigma = \{U_\alpha\}$ is an open cover of A in X, then A being compact means that there are a finite number of the U_α, denoted let's say by $U_1, \ldots, U_n$, such that

$$A \subset (U_1 \cup U_2 \cup \ldots \cup U_n).$$

It should not be at all apparent why this definition would be useful, much less important. Part of its significance will be seen in the next section when we discuss the Heine-Borel Theorem.

Definition 4.1.4 *A topological space X is* Hausdorff *if given any two points $x_1, x_2 \in X$, there are two open sets U_1, and U_2 with $x_1 \in U_1$ and $x_2 \in U_2$ but with the intersection of U_1 and U_2 empty.*

Thus X is Hausdorff if points can be isolated (separated) from each other by disjoint open sets.

Definition 4.1.5 *A function $f : X \to Y$ is* continuous, *where X and Y are two topological spaces, if given any open set U in Y, then the inverse image $f^{-1}(U)$ in X must be open.*

Definition 4.1.6 *A topological space X is* connected *if it is not possible to find two open sets U and V in X with $X = U \cup V$ and $U \cap V = \phi$.*

Definition 4.1.7 *A topological space in X is* path connected *if given any two points a and b in X, there is a continuous map*

$$f : [0,1] \to X$$

with

$$f(0) = a \text{ and } f(1) = b.$$

Here of course

$$[0,1] = \{x \in \mathbf{R} : 0 \le x \le 1\}$$

is the unit interval. To make this last definition well-defined, we would need to put a topology on this interval $[0,1]$, but this is not hard and will in fact be done in the next section.

Though in the next section the standard topology on $\mathbf{R}^n$ will be developed, we will use this topology in order to construct a topological space that is connected but is not path connected. It must be emphasized that this is a pathology. In most cases, connected is equivalent to path connected.

Let

$$X = \{(0,t) : -1 \le t \le 1\} \cup \{y = sin(\frac{1}{x}) : x > 0\}.$$

Put the induced topology on X from the standard topology on $\mathbf{R}^2$. Note that there is no path connecting the point $(0,0)$ to $(\frac{1}{\pi},0)$. In fact, no point on the segment $\{(0,t) : -1 \le t \le 1\}$ can be connected by a path to any point on the curve $\{y = sin(\frac{1}{x}) : x > 0\}$. But on the other hand, the curve $\{y = sin(\frac{1}{x}) : x > 0\}$ gets arbitrarily close to the segment $\{(0,t) : -1 \le t \le 1\}$ and hence there is no way to separate the two parts by open sets.

Point set topology books would now give many further examples of various topological spaces which satisfy some but not all of the above conditions. Most have the feel, legitimately, of pathologies, creating in some the sense that all of these definitions are somewhat pedantic and not really essential. To counter this feel, in the last section of this chapter we will look at a nonstandard topology on commutative rings, the Zariski topology, which is definitely not a pathology. But first, in the next section, we must look at the standard topology on $\mathbf{R}^n$.

4.2   The Standard Topology on $\mathbf{R}^n$

Point set topology is definitely a product of the early twentieth century. However, long before that, people were using continuous functions and related ideas. Even in previous chapters, definitions were given for continuous functions, without the need to discuss open sets and topology. In this section we define the standard topology on $\mathbf{R}^n$ and show that the definition of continuity given in the last chapter in terms of limits agrees with the definition given in the last section in terms of inverse images of open sets. The important point is that the open set version can be used in contexts for which the limit notion makes no sense. Also, in practice the open set version is frequently no harder to use than the limit version.

Critical to the definition of the standard topology on $\mathbf{R}^n$ is that there is a natural notion of distance on $\mathbf{R}^n$. Recall that the distance between two points $a = (a_1, \ldots, a_n)$ and $b = (b_1, \ldots, b_n)$ in $\mathbf{R}^n$ is defined by

$$|a - b| = \sqrt{(a_1 - b_1)^2 + \ldots + (a_n - b_n)^2}.$$

With this, we can define a topology on $\mathbf{R}^n$ by specifying as the open sets the following:

Definition 4.2.1 *A set U in $\mathbf{R}^n$ will be* open *if given any $a \in \mathbf{R}^n$, there is a real number $\epsilon > 0$ such that*

$$\{x : |x - a| < \epsilon\}$$

is contained in U.

In $\mathbf{R}^1$, sets of the form $(a, b) = \{x : a < x < b\}$ are open, while sets of the form $[a, b] = \{x : a \leq x \leq b\}$ are closed. Sets like $[a, b) = \{x : a \leq x < b\}$ are neither open nor closed. In $\mathbf{R}^2$, the set $\{(x, y) : x^2 + y^2 < 1\}$ is open.

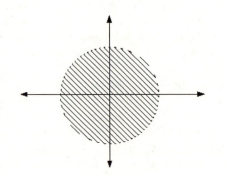

while $\{(x, y) : x^2 + y^2 \leq 1\}$ is closed.

Proposition 4.2.1 *The above definition of an open set will define a topology on* $\mathbf{R}^n$.

(The proof is exercise 2 at the end of the chapter.) This is called the standard topology on R^n.

Proposition 4.2.2 *The standard topology on* $\mathbf{R}^n$ *is Hausdorff.*

This theorem is quite obvious geometrically:

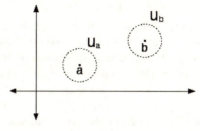

but we give a proof in order to test the definitions.

Proof: Let a and b be two distinct points in $\mathbf{R}^n$. Let $d = |a - b|$ be the distance from a to b. Set

$$U_a = \{x \in \mathbf{R}^n : |x - a| < \frac{d}{3}\}$$

and

$$U_b = \{x \in \mathbf{R}^n : |x - b| < \frac{d}{3}\}.$$

Both U_a and U_b are open sets with $a \in U_a$ and $b \in U_b$. Then $\mathbf{R}^n$ will be Hausdorff if

$$U_a \cap U_b = \phi.$$

Suppose that the intersection is not empty. Let $x \in U_a \cap U_b$. Then, by using the standard trick of adding terms that sum to zero and using the triangle inequality, we have

$$
\begin{aligned}
|a - b| &= |a - x + x - b| \\
&\leq |a - x| + |x - b| \\
&< \frac{d}{3} + \frac{d}{3} \\
&= \frac{2d}{3} \\
&< d.
\end{aligned}
$$

Since we cannot have $d = |a-b| < d$ and since the only assumption we made is that there is a point x in both U_α and U_b, we see that the intersection must indeed be empty. Hence the space $\mathbf{R}^n$ is Hausdorff. $\square$

In Chapter Three, we defined a function $f : \mathbf{R}^n \to \mathbf{R}^m$ to be continuous if, for all $a \in \mathbf{R}^n$,

$$\lim_{x \to a} f(x) = f(a),$$

meaning that given any $\epsilon > 0$, there is some $\delta > 0$ such that if $|x - a| < \delta$, then

$$|f(x) - f(a)| < \epsilon.$$

This limit definition of continuity captures much of the intuitive idea that a function is continuous if it can be graphed without lifting the pen from the page. Certainly we want this previous definition of continuity to agree with our new definition that requires the inverse image of an open set to be open. Again, the justification for the inverse image version of continuity is that it can be extended to contexts where the limit version (much less the requirement of not lifting the pen from the page) makes no sense.

Proposition 4.2.3 *Let $f : \mathbf{R}^n \to \mathbf{R}^m$ be a function. For all $a \in \mathbf{R}^n$,*

$$\lim_{x \to a} f(x) = f(a)$$

if and only, if for any open set U in $\mathbf{R}^m$, the inverse image $f^{-1}(U)$ is open in $\mathbf{R}^n$.

Proof: First assume that the inverse image of every open set in $\mathbf{R}^m$ is open in $\mathbf{R}^n$. Let $a \in \mathbf{R}^n$. We must show that

$$\lim_{x \to a} f(x) = f(a).$$

Let $\epsilon > 0$. We must find some $\delta > 0$ so that if $|x - a| < \delta$, then

$$|f(x) - f(a)| < \epsilon.$$

Define

$$U = \{y \in \mathbf{R}^m : |y - f(a)| < \epsilon\}.$$

The set U is open in $\mathbf{R}^m$. By assumption the inverse image

$$
\begin{aligned}
f^{-1}(U) &= \{x \in \mathbf{R}^n : f(x) \in U\} \\
&= \{x \in \mathbf{R}^n : |f(x) - f(a)| < \epsilon\}
\end{aligned}
$$

is open in $\mathbf{R}^n$. Since $a \in f^{-1}(U)$, there is some real number $\delta > 0$ such that the set

$$\{x : |x - a| < \delta\}$$

is contained in $f^{-1}(U)$, by the definition of open set in $\mathbf{R}^n$. But then if $|x - a| < \delta$, we have $f(x) \in U$, or in other words,

$$|f(x) - f(a)| < \epsilon,$$

which is what we wanted to show. Hence the inverse image version of continuity implies the limit version.

Now assume that

$$\lim_{x \to a} f(x) = f(a).$$

Let U be any open set in $\mathbf{R}^m$. We need to show that the inverse $f^{-1}(U)$ is open in $\mathbf{R}^n$.

If $f^{-1}(U)$ is empty, we are done, since the empty set is always open. Now assume $f^{-1}(U)$ is not empty. Let $a \in f^{-1}(U)$. Then $f(a) \in U$. Since U is open, there is a real number $\epsilon > 0$ such that the set

$$\{y \in \mathbf{R}^m : |y - f(a)| < \epsilon\}$$

is contained in the set U. Since $\lim_{x \to a} f(x) = f(a)$, by the definition of limit, given this $\epsilon > 0$, there must be some $\delta > 0$ such that if $|x - a| < \delta$, then

$$|f(x) - f(a)| < \epsilon.$$

Therefore if $|x - a| < \delta$, then $f(x) \in U$. Thus the set

$$\{x : |x - a| < \delta\}$$

is contained in the set $f^{-1}(U)$, which means that $f^{-1}(U)$ is indeed an open set. Thus the two definitions of continuity agree. $\square$

In the last section, a compact set was defined to be a set A on which every open cover $\Sigma = \{U_\alpha\}$ of A has a finite subcover. For the standard topology on $\mathbf{R}^n$, compactness is equivalent to the more intuitive idea that the set is compact if it is both closed and bounded. This equivalence is the goal of the Heine-Borel Theorem:

Theorem 4.2.1 (Heine-Borel) *A subset A of $\mathbf{R}^n$ is compact if and only if it is closed and bounded.*

We will first give a definition for boundedness, look at some examples and then sketch a proof of a special case of the theorem.

Definition 4.2.2 *A subset A is* bounded *in $\mathbf{R}^n$ if there is some fixed real number r such that for all $x \in A$,*

$$|x| < r$$

(i.e., A is contained in a ball of radius r).

For our first example, consider the open interval $(0, 1)$ in $\mathbf{R}$, which is certainly bounded, but is not closed. We want to show that this interval is also not compact. Let

$$\begin{aligned}
U_n &= \left(\frac{1}{n}, 1 - \frac{1}{n}\right) \\
&= \{x : \frac{1}{n} < x < 1 - \frac{1}{n}\}
\end{aligned}$$

be a collection of open sets.

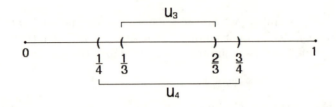

This collection will be an open cover of the interval, since every point in $(0, 1)$ is in some U_n. (In fact, once a given point is in a set U_n, it will be in every future set U_{n+k}.) But note that no finite subcollection will cover the entire interval $(0, 1)$. Thus $(0, 1)$ cannot be compact.

The next example will be of a closed but not bounded interval. Again an explicit open cover will be given for which there is no finite subcover. The interval $[0, \infty) = \{x : 0 \leq x\}$ is closed but is most definitely not bounded. It also is not compact as can be seen with the following open cover:

$$U_n = (-1, n) = \{x : -1 < x < n\}.$$

The collection $\{U_n\}_{n=1}^{\infty}$, will cover $[0, \infty)$, but can contain no finite subcover.

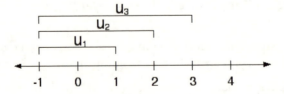

The proof of the Heine-Borel theorem revolves around reducing the whole argument to the special case of showing that a closed bounded interval on the real line is compact. (On how to reduce to this lemma, see the rigorous proof in Spivak [103], which is where we got the following argument.) This is the technical heart of the proof. The key idea actually pops up in a number of different contexts, which is why we give it here.

Lemma 4.2.1 *On the real line* $\mathbf{R}$, *a closed interval* $[a, b]$ *is compact.*

Proof: Let Σ be an open cover of $[a, b]$. We need to find a finite subcover. Define a new set

$$Y = \{x \in [a, b] : \text{there is a finite subcover in } \Sigma \text{ of the interval } [a, x]\}.$$

Our goal is to show that our interval's endpoint b is in this new set Y.

We will first show that Y is not empty, by showing that the initial point a is in Y. If $x = a$, then we are interested in the trivial interval $[a, a] = a$, a single point. Since Σ is an open cover, there is an open set $V \in \Sigma$ with $[a, a] \in V$. Thus for the admittedly silly interval $[a, a]$ there is a finite subcover, and thus a is in the set Y, meaning that, at the least, Y is not empty.

Set α to be the least upper bound of Y. This means that there are elements in Y arbitrarily close to α but that no element of Y is greater than α. (Though to show the existence of such a least upper bound involves the subtle and important property of completeness of the real number line, it is certainly quite reasonable intuitively that such an upper bound must exist for any bounded set of reals.) We first show that the point α is itself in the set Y and, second, that α is in fact the endpoint b, which will allow us to conclude that the interval is indeed compact.

Since $\alpha \in [a, b]$ and since Σ is an open cover, there is an open set U in Σ with $\alpha \in U$. Since U is open in $[a, b]$, there is a positive number ϵ with

$$\{x : |x - \alpha| < \epsilon\} \subset U.$$

Since α is the least upper bound of Y, there must be an $x \in Y$ that is arbitrarily close to but less than α. Thus we can find an $x \in Y \cap U$ with

$$\alpha - x < \epsilon,$$

Since $x \in Y$, there is a finite subcover $U_1, \ldots, U_N$ of the interval $[a, x]$. Then the finite collection $U_1, \ldots, U_N, U$ will cover $[a, \alpha]$. But this means, since each open set U_k and U are in Σ, that the interval $[a, \alpha]$ has a finite subcover and hence that the least upper bound α is in Y.

Now assume $\alpha < b$. We want to come up with a contradiction. We know that α is in the set Y. Hence there is a finite subcover $U_1, \ldots, U_n$ of the collection Σ which will cover the interval $[a, \alpha]$. Choose the open sets so that the point α is in the open set U_n. Since U_n is open, there is an $\epsilon > 0$ with

$$\{x : |x - \alpha| < \epsilon\} \subset U_n.$$

Since the endpoint b is strictly greater than the point α, we can actually find a point x that both is in the open set U_n and satisfies

$$\alpha < x < b.$$

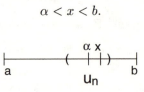

But then the finite subcover $U_1, \ldots, U_n$ will cover not only the interval $[a, \alpha]$ but also the larger interval $[a, x]$, forcing the point x to be in the set Y. This is impossible, since α is the largest possible element in Y. Since the only assumption that we made was that $\alpha < b$, we must have $\alpha = b$, as desired. □

There is yet another useful formulation for compactness in $\mathbf{R}^n$.

Theorem 4.2.2 *A subset A in $\mathbf{R}^n$ is compact if every infinite sequence (x_n) of points in A has a subsequence converging to a point in A. Thus, if (x_n) is a collection of points in A, there must be a point $p \in A$ and a subsequence x_{n_k} with $\lim_{k \to \infty} x_{n_k} = p$.*

The proof is one of the exercises at the end of the chapter.

Compactness is also critical for the following:

Theorem 4.2.3 *Let X be a compact topological space and let $f : X \to \mathbf{R}$ be a continuous function. Then there is a point $p \in X$ where f has a maximum.*

We give a general idea of the proof, with the details saved for the exercises. First, we need to show that the continuous image of a compact set is compact. Then $f(X)$ will be compact in $\mathbf{R}$ and hence must be closed and bounded. Thus there will be a least upper bound in $f(X)$, whose inverse image will contain the desired point p. A similar argument can be used to show that any continuous function $f(x)$ on a compact set must also have a minimum.

4.3   Metric Spaces

The natural notion of distance on the set $\mathbf{R}^n$ is the key to the existence of the standard topology. Luckily on many other sets similar notions of distance (called metrics) exist; any set that has a metric automatically has a topology.

Definition 4.3.1 *A* metric *on a set X is a function*

$$\rho : X \times X \to \mathbf{R}$$

such that for all points $x, y, z \in X$ we have:

1. *$\rho(x, y) \geq 0$ and $\rho(x, y) = 0$ if and only if $x = y$.*

2. *$\rho(x, y) = \rho(y, x)$.*

3. *(Triangle Inequality)*

$$\rho(x, z) \leq \rho(x, y) + \rho(y, z).$$

The set X with its metric ρ is called a metric space *and is denoted by (X, ρ).*

Fix a metric space (X, ρ).

Definition 4.3.2 *A set U in X is* open *if for all points $a \in U$, there is some real number $\epsilon > 0$ such that*

$$\{x : |x - a| < \epsilon\}$$

is contained in U.

Proposition 4.3.1 *The above definition for open set will define a Hausdorff topological space on the metric space (X, ρ).*

The proof is similar to the corresponding proof for the standard topology on $\mathbf{R}^n$. In fact, most of the topological facts about $\mathbf{R}^n$ can be quite easily translated into corresponding topological facts about any metric space. Unfortunately, as will be seen in section five, not all natural topological spaces come from a metric.

An example of a metric that is not just the standard one on $\mathbf{R}^n$ is given in Chapter Thirteen, when a metric and its associated topology is used to define Hilbert spaces.

4.4 Bases for Topologies

Warning: This section uses the notion of countability. A set is countable if there is a one-to-one onto mapping from the set to the natural numbers. More on this is in Chapter Ten. Note that the rational numbers are countable while the real numbers are uncountable.

In linear algebra, the word *basis* means a list of vectors in a vector space that generates uniquely the entire vector space. In a topology, a basis will be a collection of open sets that generate the entire topology. More precisely:

Definition 4.4.1 *Let X be a topological space. A collection of open sets forms a* basis *for the topology if every open set in X is the (possibly infinite) union of sets from the collection.*

For example, let (X, ρ) be a metric space. For each positive integer k and for each point $p \in X$, set

$$U(p, k) = \{x \in X : \rho(x, p) < \frac{1}{k}\}.$$

We can show that the collection of all possible $U(p, k)$ forms a basis for the topology of the metric space.

In practice, having a basis will allow us to reduce many topological calculations to calculating on sets in the basis. This will be more tractable if we can somehow limit the number of elements in a basis. This leads to

Definition 4.4.2 *A topological space is* second countable *if it has a basis with a countable number of elements.*

For example, $\mathbf{R}^n$, with the usual topology, is second countable. A countable basis can be constructed as follows. For each positive integer k and each $p \in \mathbf{Q}^n$ (which means that each coordinate of the point p is a rational number), define

$$U(p, k) = \{x \in \mathbf{R}^n : |x - p| < \frac{1}{k}\}.$$

There are a countable number of such sets $U(p, k)$ and they can be shown to form a basis.

Most reasonable topological spaces are second countable. Here is an example of a metric space that is not second countable. It should and does have the feel of being a pathology. Let X be any uncountable set (you can, for example, let X be the real numbers). Define a metric on X by setting $\rho(x, y) = 1$ if $x \neq y$ and $\rho(x, x) = 0$. It can be shown that this ρ defines a metric on X and thus defines a topology on X. This topology is weird, though. Each point x is itself an open set, since the open set $\{y \in X : \rho(x, y) < 1/2\} = x$. By using the fact that there are an uncountable number of points in X, we can show that this metric space is not second countable.

Of course, if we use the term "second countable", there must be a meaning to "first countable". A topological set is *first countable* if every point $x \in X$ has a countable neighborhood basis. For this to make sense, we need to know what a neighborhood basis is. A collection of open sets in X forms a *neighborhood basis* of some $x \in X$ if every open set containing x has in it an open set from the collection and if each open set in the collection contains the point x. We are just mentioning this definition for the sake of completeness. While we will later need the notion of second countable, we will not need in this book the idea of first countable.

4.5 Zariski Topology of Commutative Rings

Warning: This section requires a basic knowledge of commutative ring theory.

Though historically topology arose in the study of continuous functions on $\mathbf{R}^n$, a major reason why all mathematicians can speak the language of open, closed and compact sets is because there exists natural topologies on many diverse mathematical structures. This section looks at just one of these topologies. While this example (the Zariski topology for commutative rings) is important in algebraic geometry and algebraic number theory, there is no reason for the average mathematician to know it. It is given here simply to show how basic topological notions can be applied in a nonobvious way to an object besides $\mathbf{R}^n$. We will in fact see that the Zariski topology on the ring of polynomials is not Hausdorff and hence cannot come from a metric.

We want to associate a topological space to any commutative ring R. Our topological space will be defined on the set of all prime ideals in the ring R, a set that will be denoted by $Spec(R)$. Instead of first defining the open sets, we will start with what will be the closed sets. Let $\mathcal{P}$ be a prime ideal in R and hence a point in Spec R. Define *closed* sets to be

$$V_{\mathcal{P}} = \{\mathcal{Q} : \mathcal{Q} \text{ is a prime ideal in } \mathcal{R} \text{ containing } \mathcal{P}\}.$$

Then define Spec $R - V_{\mathcal{P}}$, where $\mathcal{P}$ is any prime ideal, to be an open set. The *Zariski topology* on Spec R is given by defining open sets to be the unions and finite intersections of all sets of the form Spec $R - V_{\mathcal{P}}$.

As will be seen in some of the examples, it is natural to call the points in Spec R corresponding to maximal ideals *geometric points*.

Assume that the ring R has no zero divisors, meaning that if $x \cdot y = 0$, then either x or y must be zero. Then the element 0 will generate a prime ideal, (0), contained in every other ideal. This ideal is called the *generic ideal* and is always a bit exceptional.

Now for some examples. For the first, let the ring R be the integers $\mathbf{Z}$. The only prime ideals in $\mathbf{Z}$ are of the form

$$(p) = \{kp : k \in \mathbf{Z}, p \text{ a prime number}\}$$

and the zero ideal (0). Then Spec $\mathbf{Z}$ is the set of all prime numbers:

and the zero ideal (0). The open sets in this topology are the complements of a finite number of these ideals.

For our second example, let the ring R be the field of complex numbers $\mathbf{C}$. The only two prime ideals are the zero ideal (0) and the whole field itself. Thus in some sense the space $\mathbf{C}$ is a single point.

A more interesting example occurs by setting $R = \mathbf{C}[x]$, the ring of one-variable polynomials with complex coefficients. We will see that as a point set this space can be identified with the real plane $\mathbf{R}^2$ (if we do not consider the generic ideal) but that the topology is far from the standard topology of $\mathbf{R}^2$. Key is that all one-variable polynomials can be factored into linear factors, by the Fundamental Theorem of Algebra; thus all prime ideals are multiples of linear polynomials. We denote the ideal of all of the multiples of a linear polynomial $x - c$ as:

$$(x - c) = \{f(x)(x - c) : f(x) \in \mathbf{C}[x], c \in \mathbf{C}\}.$$

Hence, to each complex number, $c = a + bi$ with $a, b \in \mathbf{R}$, there corresponds a prime ideal $(x - c)$ and thus Spec $\mathbf{C}[x]$ is another, more ring-theoretic description of the complex numbers. Geometrically, Spec $\mathbf{C}[x]$ is

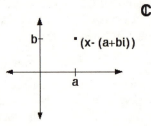

Note that while the zero ideal (0) is still a prime ideal in $\mathbf{C}[x]$, it does not correspond to any point in $\mathbf{C}$; instead, it is lurking in the background. The open sets in this topology are the complements of a finite number of the prime ideals. But each prime ideal corresponds to a complex number. Since the complex numbers $\mathbf{C}$ can be viewed as the real plane $\mathbf{R}^2$, we have that an open set is the complement of a finite number of points in the real plane. While these open sets are also open in the standard topology on $\mathbf{R}^2$, they are far larger than any open disc in the plane. No little ϵ-disc will be the complement of only a finite number of points and hence cannot be open in the Zariski topology. In fact, notice that the intersection of two of these Zariski open sets must intersect. This topology cannot be Hausdorff. Since all metric spaces are Hausdorff, this means that the Zariski topology cannot come from some metric.

Now let $R = \mathbf{C}[x, y]$ be the ring of two-variable polynomials with complex coefficients. Besides the zero ideal (0), there are two types of prime ideals: the maximal ideals, each of which is generated by polynomials of the form $x - c$ and $y - d$, where c and d are any two complex numbers and nonmaximal prime ideals, each of which is generated by an irreducible polynomial $f(x, y)$.

Note that the maximal ideals correspond to points in the complex plane $\mathbf{C} \times \mathbf{C}$, thus justifying the term 'geometric point'.

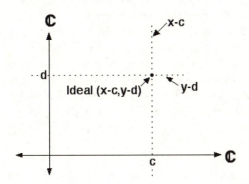

Since each copy of the complex numbers $\mathbf{C}$ is a real plane $\mathbf{R}^2$, $\mathbf{C} \times \mathbf{C}$ is $\mathbf{R}^2 \times \mathbf{R}^2 = \mathbf{R}^4$. In the Zariski topology, open sets are the complements of the zero loci of polynomials. For example, if $f(x, y)$ is an irreducible polynomial, then the set

$$U = \{(x, y) \in \mathbf{C}^2 : f(x, y) \neq 0\}$$

is open. While Zariski sets will still be open in the standard topology on $\mathbf{R}^4$, the converse is most spectacularly false. Similar to the Zariski topology on $\mathbf{C}[x]$, no ϵ-ball will be open in the Zariski topology on $\mathbf{C}[x, y]$. In fact, if U and V are two Zariski open sets that are non-empty, they must intersect. Thus this is also a non-Hausdorff space and hence cannot come from a metric space.

4.6 Books

Point set topology's days of glory were the early twentieth century, a time when some of the world's best mathematicians were concerned with the correct definitions for continuity, dimension and for a topological space. Most of these issues have long been settled. Today, point set topology is overwhelmingly a tool that all mathematicians need to know.

At the undergraduate level, it is not uncommon for a math department to use their point set topology class as a place to introduce students to proofs. Under the influence of E. H. Moore (of the University of Chicago) and of his student R.L. Moore (of the University of Texas, who advised an amazing number of Ph.D. students), many schools have taught topology under the Moore method. Using this approach, on the first day of class students are given a list of the definitions and theorems. On the second day people are asked who has proven Theorem One. If someone thinks they have a proof, they go to the board to present it to the class. Those who still want to think of a proof on their own leave the class for that part of the lecture. This is a powerful way to introduce students to proofs. On the other hand, not much material can be covered. At present, most people who teach using the Moore method modify it in various ways.

Of course, this approach comes close to being absurd for people who are already mathematically mature and just need to be able to use the results. The texts of the fifties and sixties were by Kelley [72] and Dugundji [30]. Overwhelmingly the most popular current book is Munkres' *Topology: A First Course* [88].

My own bias (a bias not shared by most) is that all the point set topology that most people need can be found in, for example, the chapter in Royden's *Real Analysis* [95] on topology.

4.7   Exercises

1. The goal of this problem is to show that a topology on a set X can also be defined in terms of a collection of closed sets, as opposed to a collection of open sets. Let X be a set of points and let $C = \{C_\alpha\}$ be a collection of subsets of X. Suppose that

- Any finite union of sets in the collection C must be another set in C.

- Any intersection of sets in C must be another set in C.

- The empty set ϕ and the whole space X must in the collection C.

Call the sets in C *closed* and call a set U *open* if its complement $X - U$ is closed. Show that this definition of open set will define a topology on the set X.

2. Prove Proposition 4.2.1.

3. Prove Theorem 4.2.2.

4. Prove Theorem 4.2.3.

5. Let V be the vector space of all functions

$$f : [0, 1] \to \mathbf{R}$$

whose derivatives, including the one-sided derivatives at the endpoints, are continuous functions on the interval $[0, 1]$. Define

$$|f|_\infty = \sup_{x \in [0,1]} |f(x)|$$

for any function $f \in V$. For each $f \in V$ and each $\epsilon > 0$, define

$$U_f(\epsilon) = \{g \in V : |f - g|_\infty < \epsilon\}.$$

    a. Show that the set of all $U_f(\epsilon)$ is a basis for a topology on the set V.

    b. Show that there can be no number M such that for all $f \in V$,

$$|\frac{\mathrm{d}f}{\mathrm{d}x}|_\infty < M|f|_\infty.$$

In the language of functional analysis, this means that the derivative, viewed as a linear map, is not *bounded* on the space V. One of the main places where serious issues involving point set topology occur is in functional analysis, which is the study of vector spaces of various types of functions. The study of such space is important in trying to solve differential equations.

Chapter 5

Classical Stokes' Theorems

Basic Objects:	Manifolds and boundaries
Basic Maps:	Vector-valued functions on manifolds
Basic Goal:	Function's average over a boundary
	= Derivative's average over interior

Stokes' Theorem, in all of its many manifestations, comes down to equating the average of a function on the boundary of some geometric object with the average of its derivative (in a suitable sense) on the interior of the object. Of course, a correct statement about averages must be put into the language of integrals. This theorem provides a deep link between topology (the part about boundaries) and analysis (integrals and derivatives). It is also critical for much of physics, as can be seen in both its historical development and in the fact that for most people their first introduction to Stokes' Theorem is in a course on electricity and magnetism.

The goal of Chapter Six is to prove Stokes' Theorem for abstract manifolds (which are, in some sense, the abstract method for dealing with geometric objects). As will be seen, to even state this theorem takes serious work in building up the necessary machinery. This chapter looks at some special cases of Stokes' Theorem, special cases that were known long before people realized that there is this one general underlying theorem. For example, we will see that the Fundamental Theorem of Calculus is a special case of Stokes' Theorem (though to prove Stokes' Theorem, you use the Fundamental Theorem of Calculus; thus logically Stokes' Theorem does not imply the Fundamental Theorem of Calculus). It was in the 1800s that most of these special cases of Stokes' Theorem were discovered, though, again, people did not know that each of these were special cases of one

general result. These special cases are important and useful enough that
they are now standard topics in most multivariable calculus courses and
introductory classes in electricity and magnetism. They are Green's Theorem, the Divergence Theorem and Stokes' Theorem. (This Stokes' theorem
is, though, a special case of the Stokes' Theorem of the next chapter.) This
chapter develops the needed mathematics for these special cases. We will
state and sketch proofs for the Divergence Theorem and Stokes' Theorem.
Physical intuitions will be stressed.

There is a great deal of overlap between the next chapter and this one.
Mathematicians need to know both the concrete special cases of Stokes'
Theorem and the abstract version of Chapter Six.

5.1   Preliminaries about Vector Calculus

This is a long section setting up the basic definitions of vector calculus. We
need to define vector fields, manifolds, path and surface integrals, divergence and curl. All of these notions are essential. Only then can we state
the Divergence Theorem and Stokes' Theorem, which are the goals of this
chapter.

5.1.1   Vector Fields

Definition 5.1.1 *A vector field on* $\mathbf{R}^n$ *is a vector-valued function*

$$\mathbf{F} : \mathbf{R}^n \to \mathbf{R}^m.$$

If $x_1, \ldots, x_n$ *are coordinates for* $\mathbf{R}^n$, *then the vector field* $\mathbf{F}$ *will be described
by* m *real-valued functions* $f_k : \mathbf{R}^n \to \mathbf{R}$ *as follows:*

$$\mathbf{F}(x_1, \ldots, x_n) = \begin{pmatrix} f_1(x_1, \ldots, x_n) \\ \vdots \\ f_m(x_1, \ldots, x_n) \end{pmatrix}.$$

A vector field is *continuous* if each real-valued function f_k is continuous,
differentiable if each real-valued f_k is differentiable, etc.

Intuitively, a vector field assigns to each point of $\mathbf{R}^n$ a vector. Any
number of physical phenomenon can be captured in terms of vector fields.
In fact, they are the natural language of fluid flow, electric fields, magnetic
fields, gravitational fields, heat flow, traffic flow and much more.

For example, let $\mathbf{F} : \mathbf{R}^2 \to \mathbf{R}^2$ be given by

$$\mathbf{F}(x, y) = (3, 1).$$

Here $f_1(x, y) = 3$ and $f_2(x, y) = 1$. On $\mathbf{R}^2$ this vector field can be pictured by drawing in a few sample vectors.

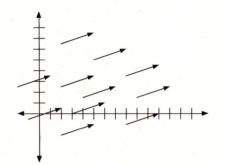

A physical example of this vector field would be wind blowing in the direction $(3, 1)$ with velocity

$$\text{length}(3, 1) = \sqrt{9 + 1} = \sqrt{10}.$$

Now consider the vector field $\mathbf{F}(x, y) = (x, y)$. Then in pictures we have:

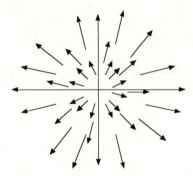

This could represent water flowing out from the origin $(0, 0)$.

For our final example, let $\mathbf{F}(x, y) = (-y, x)$. In pictures we have:

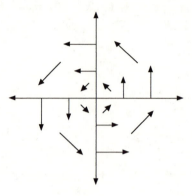

which might be some type of whirlpool.

5.1.2   Manifolds and Boundaries

Curves and surfaces appear all about us. Both are examples of manifolds, which are basically just certain naturally occurring geometric objects. The intuitive idea of a manifold is that, for a k-dimensional manifold, each point is in a neighborhood that looks like a ball in $\mathbf{R}^k$. In the next chapter we give three different ways for defining a manifold. In this chapter, we will define manifolds via *parametrizations*. The following definition is making rigorous the idea that locally, near any point, a k-dimensional manifold looks like a ball in $\mathbf{R}^k$.

Definition 5.1.2 *A differentiable manifold M of dimension k in $\mathbf{R}^n$ is a set of points in $\mathbf{R}^n$ such that for any point $p \in M$, there is a small open neighborhood U of p, a vector-valued differentiable function $F : \mathbf{R}^k \to \mathbf{R}^n$ and an open set V in $\mathbf{R}^k$ with*
  a) $F(V) = U \cap M$

  b) The Jacobian of F has rank k at every point in V, where the Jacobian *of F is the $n \times k$ matrix*

$$\begin{pmatrix} \frac{\partial f_1}{\partial x_1} & \cdots & \frac{\partial f_1}{\partial x_k} \\ \vdots & & \vdots \\ \frac{\partial f_n}{\partial x_1} & \cdots & \frac{\partial f_n}{\partial x_k} \end{pmatrix},$$

with $x_1, \ldots, x_k$ a coordinate system for $\mathbf{R}^\mathbf{k}$. The function F is called the (local) parametrization *of the manifold.*

Recall that the rank of a matrix is k if the matrix has an invertible $k \times k$ minor. (A minor is a submatrix of a matrix.)

A circle is a one-dimensional manifold, with a parametrization

$$F : \mathbf{R}^1 \to \mathbf{R}^2$$

given by

$$F(t) = (\cos(t), \sin(t)).$$

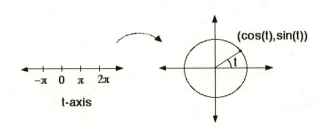

Geometrically the parameter t is the angle with the x-axis. Note that the Jacobian of F is $\begin{pmatrix} -\sin t \\ \cos t \end{pmatrix}$. Since *sine* and *cosine* cannot simultaneously be zero, the Jacobian has rank 1.

A cone in three-space can be parametrized by

$$F(u, v) = (u, v, \sqrt{u^2 + v^2}).$$

This will be a two dimensional manifold (a surface) except at the vertex $(0, 0, 0)$, for at this point the Jacobian fails to be well-defined, much less having rank two. Note that this agrees with the picture, where certainly the origin looks quite different than the other points.

Again, other definitions are given in Chapter Six.

Now to discuss what is the boundary of a manifold. This is needed since Stokes' Theorem and its many manifestations state that the average of a function on the boundary of a manifold will equal the average of its derivative on the interior.

Let M be a k-dimensional manifold in $\mathbf{R}^n$.

Definition 5.1.3 *The* closure *of M, denoted $\bar{M}$, is the set of all points x in $\mathbf{R}^n$ such that there is a sequence of points (x_n) in the manifold M with*

$$\lim_{n \to \infty} x_n = x.$$

The boundary *of M, denoted ∂M, is:*

$$\partial M = \bar{M} - M,$$

Given a manifold with boundary, we call the nonboundary part the *interior*.

All of this will become relatively straightforward with a few examples. Consider the map

$$r : [-1, 2] \to \mathbf{R}^2$$

where

$$r(t) = (t, t^2).$$

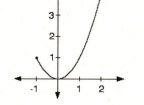

The image under r of the open interval $(-1, 2)$ is a one-manifold (since the Jacobian is the 2×1 matrix $(1, 2t)$, which always has rank one). The boundary consists of the two points $r(-1) = (-1, 1)$ and $r(2) = (2, 4)$.

Our next example is a two-manifold having a boundary consisting of a circle. Let

$$r : \{(x, y) \in \mathbf{R}^2 : x^2 + y^2 \leq 1\} \to \mathbf{R}^3$$

be defined by

$$r(x, y) = (x, y, x^2 + y^2).$$

The image of r is a bowl in space sitting over the unit disc in the plane:

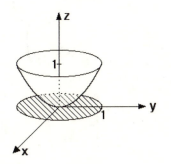

Now the image under r of the open disc $\{(x,y) \in \mathbf{R}^2 : x^2 + y^2 < 1\}$ is a two-manifold (since the Jacobian is

$$\begin{pmatrix} 1 & 0 & 2x \\ 0 & 1 & 2y \end{pmatrix},$$

which has rank two at all points). The boundary is the image of the boundary of the disc and hence the image of the circle $\{(x,y) \in \mathbf{R}^2 : x^2 + y^2 = 1\}$. In this case, as can be seen by the picture, the boundary is itself a circle living on the plane $z = 1$ in space.

Another example is the unit circle in the plane. We saw that this is a one-manifold. There are no boundary points, though. On the other hand, the unit circle is itself the boundary of a two-manifold, namely the unit disc in the plane. In a similar fashion, the unit sphere in $\mathbf{R}^3$ is a two-manifold, with no boundary, that is itself the boundary of the unit ball, a three-manifold. (It is not chance that in these two cases that the boundary of the boundary is the empty set.)

We will frequently call a manifold with boundary simply a manifold. We will also usually be making the assumption that the boundary of an n-dimensional manifold will either be empty (in which case the manifold has no boundary) or is itself an $(n-1)$-dimensional manifold.

5.1.3   Path Integrals

Now that we have a sharp definition for manifolds, we want to do calculus on them. We start with integrating vector fields along curves. This process is called a path integral or sometimes, misleadingly, a line integral.

A *curve* or *path* C in $\mathbf{R}^n$ is defined to be a one-manifold with boundary. Thus all curves are defined by maps $F : [a,b] \to \mathbf{R}^n$, given by

$$F(t) = \begin{pmatrix} f_1(t) \\ \vdots \\ f_n(t) \end{pmatrix}.$$

These maps are frequently written as

$$\begin{pmatrix} x_1(t) \\ \vdots \\ x_n(t) \end{pmatrix}.$$

We will require each component function $f_i : \mathbf{R} \to \mathbf{R}$ to be differentiable.

Definition 5.1.4 *Let* $f(x_1, \ldots, x_n)$ *be a real-valued function defined on* $\mathbf{R}^n$. *The* path integral *of the function* f *along the curve* C *is*

$$\int_c f \, ds \;=\; \int_c f(x_1, \ldots, x_n) ds$$

$$=\; \int_a^b f(x_1(t), \ldots, x_n(t)) \left(\sqrt{(\frac{dx_1}{dt})^2 + \cdots + (\frac{dx_n}{dt})^2} \right) dt.$$

Note that

$$\int_a^b f(x_1(t), \ldots, x_n(t)) \left(\sqrt{(\frac{dx_1}{dt})^2 + \cdots + (\frac{dx_n}{dt})^2} \right) dt,$$

while looking quite messy, is an integral of the single variable t.

Theorem 5.1.1 *Let a curve* C *in* $\mathbf{R}^n$ *be described by two different parametrizations*

$$F : [a, b] \to \mathbf{R}^n$$

and

$$G : [c, d] \to \mathbf{R}^n,$$

with $F(t) = \begin{pmatrix} x_1(t) \\ \vdots \\ x_n(t) \end{pmatrix}$ *and* $G(u) = \begin{pmatrix} y_1(u) \\ \vdots \\ y_n(u) \end{pmatrix}.$

The path integral $\int_C f \, ds$ *is independent of parametrization chosen, i.e.,*

$$\int_a^b f(x_1(t), \ldots, x_n(t)) \sqrt{(\frac{dx_1}{dt})^2 + \cdots + (\frac{dx_n}{dt})^2} \, dt$$

$$=$$

$$\int_c^d f(y_1(u), \ldots, y_n(u)) \sqrt{(\frac{dy_1}{du})^2 + \cdots + (\frac{dy_n}{du})^2} \, du.$$

While we will do an example in a moment, the proof uses critically and is an exercise in the chain rule. In fact, the path integral was defined with the awkward term

$$ds = \sqrt{(\frac{dx_1}{dt})^2 + \cdots + (\frac{dx_n}{dt})^2}\ dt$$

precisely in order to make the path integral independent of parametrization. This is why $\int_a^b f(x_1(t), \ldots, x_n(t))\ dt$ is an incorrect definition for the path integral.

The symbol "ds" represents the infinitesimal arc length on the curve C in $\mathbf{R}^n$. In pictures, for $\mathbf{R}^2$, consider the following.

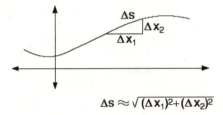

$$\Delta s \approx \sqrt{(\Delta x_1)^2 + (\Delta x_2)^2}$$

With Δs denoting the change in position along the curve C, we have by the Pythagorean Theorem

$$\Delta s \approx \sqrt{(\Delta x_1)^2 + (\Delta x_2)^2}$$
$$= \left(\sqrt{(\frac{\Delta x_1}{\Delta t})^2 + (\frac{\Delta x_2}{\Delta t})^2} \right)\ \Delta t.$$

Then in the limit as $\Delta t \to 0$, we have, at least formally,

$$ds = \left(\sqrt{(\frac{dx_1}{dt})^2 + (\frac{dx_2}{dt})^2} \right)\ dt.$$

Thus the correct implementation of the Pythagorean Theorem will also force on us the term $ds = \sqrt{(\frac{dx_1}{dt})^2 + \cdots + (\frac{dx_n}{dt})^2}\ dt$ in the definition of the path integral.

Now for an example, in order to check our working knowledge of the definitions and also to see how the ds term is needed to make path integrals independent of parametrizations. Consider the straight line segment in the plane from $(0,0)$ to $(1,2)$. We will parametrize this line segment in two different ways, and then compute the path integral of the function

$$f(x,y) = x^2 + 3y$$

using each of the parametrizations.

First, define

$$F : [0,1] \to \mathbf{R}^2$$

by

$$F(t) = (t, 2t).$$

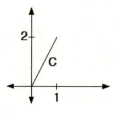

Thus we have $x(t) = t$ and $y(t) = 2t$. Denote this line segment by C. Then

$$
\begin{aligned}
\int_C f(x,y)ds &= \int_0^1 (x(t)^2 + 3y(t))\sqrt{(\frac{dx}{dt})^2 + (\frac{dy}{dt})^2}\ dt \\
&= \int_0^1 (t^2 + 6t)\sqrt{5}\ dt \\
&= \sqrt{5}(\frac{t^3}{3}\,|_0^1 + 3t^2\,|_0^1) \\
&= \sqrt{5}(\frac{1}{3} + 3) \\
&= \frac{10}{3}\sqrt{5}.
\end{aligned}
$$

Now parametrize the segment C by:

$$G : [0,2] \to C$$

where

$$G(t) = (\frac{t}{2}, t).$$

Here we have $x(t) = \frac{t}{2}$ and $y(t) = t$. Then

$$
\begin{aligned}
\int_C f(x, y) ds &= \int_0^2 (x(t)^2 + 3y(t)) \sqrt{(\frac{dx}{dt})^2 + (\frac{dy}{dt})^2} \, dt \\
&= \int_0^2 (\frac{t^2}{4} + 3t) \sqrt{\frac{1}{4} + 1} \, dt \\
&= \frac{\sqrt{5}}{2} (\frac{t^3}{12} |_0^2 + \frac{3t^2}{2} |_0^2) \\
&= \frac{\sqrt{5}}{2} (\frac{8}{12} + 6) \\
&= \frac{10}{3} \sqrt{5},
\end{aligned}
$$

as desired.

5.1.4 Surface Integrals

Now to integrate along surfaces. A *surface* in $\mathbf{R}^3$ is a two-manifold with boundary. For the sake of simplicity, we will restrict our attention to those surfaces which are the image of a map

$$ r : D \to \mathbf{R}^3, $$

given by

$$ r(u, v) = (x(u, v), y(u, v), z(u, v)), $$

where x, y, z are coordinates for $\mathbf{R}^3$ and u, v are coordinates for $\mathbf{R}^2$. Here D is a *domain* in the plane, which means that there is an open set U in $\mathbf{R}^2$ whose closure is D. (If you think of U as an open disc and D as a closed disc, you usually will not go wrong.)

Definition 5.1.5 *Let* $f(x, y, z)$ *be a function on* $\mathbf{R}^3$. *Then the integral of* $f(x, y, z)$ *along the surface* S *is*

$$ \int \int_S f(x, y, z) dS = \int \int_D f(x(u, v), y(u, v), z(u, v)) \cdot \left| \frac{\partial r}{\partial u} \times \frac{\partial r}{\partial v} \right| du dv. $$

Here $|\frac{\partial r}{\partial u} \times \frac{\partial r}{\partial v}|$ denotes the length of the cross product (which in a moment we will show to be the length of a certain normal vector) of the vectors $\frac{\partial r}{\partial u}$ and $\frac{\partial r}{\partial v}$, and is hence the determinant of

$$ \frac{\partial r}{\partial u} \times \frac{\partial r}{\partial v} = \begin{pmatrix} \mathbf{i} & \mathbf{j} & \mathbf{k} \\ \partial x/\partial u & \partial y/\partial u & \partial z/\partial u \\ \partial x/\partial v & \partial y/\partial v & \partial z/\partial v \end{pmatrix}. $$

Thus the *infinitesimal area* dS is:

$$\text{length of }(\frac{\partial r}{\partial u} \times \frac{\partial r}{\partial v})dudv =$$

$$\left| \left((\frac{\partial y}{\partial u}\frac{\partial z}{\partial v}) - (\frac{\partial z}{\partial u}\frac{\partial y}{\partial v}), (\frac{\partial x}{\partial v}\frac{\partial z}{\partial u}) - (\frac{\partial x}{\partial u}\frac{\partial z}{\partial v}), (\frac{\partial x}{\partial u}\frac{\partial y}{\partial v}) - (\frac{\partial x}{\partial v}\frac{\partial y}{\partial u}) \right) \right| dudv.$$

In analogy with arc length, a surface integral is independent of parametrization:

Theorem 5.1.2 *The integral $\int \int_S f(x, y, z) \, dS$ is independent of the parametrization of the surface S.*

Again, the chain rule is a critical part of the proof.

Note that if this theorem were not true, we would define the surface integral (in particular the infinitesimal area) differently.

We now show how the vector field

$$\frac{\partial r}{\partial u} \times \frac{\partial r}{\partial v}$$

is actually a normal to the surface. With the map $r : \mathbf{R}^2 \to \mathbf{R}^3$ given by $r(u, v) = (x(u, v), y(u, v), z(u, v))$, recall that the Jacobian of r is

$$\begin{pmatrix} \partial x/\partial u & \partial x/\partial v \\ \partial y/\partial u & \partial y/\partial v \\ \partial z/\partial u & \partial z/\partial v \end{pmatrix}.$$

But as we saw in Chapter Three, the Jacobian maps tangent vectors to tangent vectors. Thus the two vectors

$$(\frac{\partial x}{\partial u}, \frac{\partial y}{\partial u}, \frac{\partial z}{\partial u})$$

and

$$(\frac{\partial x}{\partial v}, \frac{\partial y}{\partial v}, \frac{\partial z}{\partial v})$$

are both tangent vectors to the surface S. Hence their cross product must be a normal (perpendicular) vector $\mathbf{n}$. Thus we can interpret the surface integral as

$$\int \int_S f \, dS = \int \int f \cdot |\mathbf{n}| \, dudv$$

with dS =(length of the normal vector $\frac{\partial r}{\partial u} \times \frac{\partial r}{\partial v}$) $dudv$.

5.1.5 The Gradient

The gradient of a function can be viewed as a method for differentiating functions.

Definition 5.1.6 *The* gradient *of a real-valued function* $f(x_1, \ldots, x_n)$ *is*

$$\bigtriangledown f = (\frac{\partial f}{\partial x_1}, \ldots, \frac{\partial f}{\partial x_n}).$$

Thus

$$\bigtriangledown : (\text{Functions}) \to (\text{Vector fields}).$$

For example, if $f(x, y, z) = x^3 + 2xy + 3xz$, we have

$$\bigtriangledown(f) = (3x^2 + 2y + 3z, 2x, 3x).$$

It can be shown that if at all points on $M = (f(x_1, \ldots, x_n) = 0)$ where $\bigtriangledown f \neq 0$, the gradient $\bigtriangledown f$ is a normal vector to M.

5.1.6 The Divergence

The divergence of a vector field can be viewed as a reasonable way to differentiate a vector field. (In the next section we will see that the curl of a vector field is another way.) Let $\mathbf{F}(x, y, z) : \mathbf{R}^3 \to \mathbf{R}^3$ be a vector field given by three functions as follows:

$$\mathbf{F}(x, y, z) = (f_1(x, y, z), f_2(x, y, z), f_3(x, y, z)).$$

Definition 5.1.7 *The* divergence *of* $\mathbf{F}(x, y, z)$ *is*

$$div(\mathbf{F}) = \frac{\partial f_1}{\partial x} + \frac{\partial f_2}{\partial y} + \frac{\partial f_3}{\partial z}.$$

Thus

$$\text{div} : (\text{Vector fields}) \to (\text{Functions}).$$

The Divergence Theorem will tell us that the divergence measures how much the vector field is spreading out at a point.

For example, let $\mathbf{F}(x, y, z) = (x, y^2, 0)$. Then

$$\text{div}(\mathbf{F}) = \frac{\partial x}{\partial x} + \frac{\partial(y^2)}{\partial y} + \frac{\partial(0)}{\partial z} = 1 + 2y.$$

If you sketch out this vector field, you do indeed see that the larger the y value, the more spread out the vector field becomes.

5.1.7   The Curl

The curl of a vector field is another way in which we can extend the idea of differentiation to vector fields. Stokes' Theorem will show us that the curl of a vector field measures how much the vector field is twirling or whirling or curling about. The actual definition is:

Definition 5.1.8 *The* curl *of a vector field* $\mathbf{F}(x, y, z)$ *is*

$$
\begin{aligned}
curl(\mathbf{F}) \quad &= \quad \det \begin{pmatrix} \mathbf{i} & \mathbf{j} & \mathbf{k} \\ \frac{\partial}{\partial x} & \frac{\partial}{\partial y} & \frac{\partial}{\partial z} \\ f_1 & f_2 & f_3 \end{pmatrix} \\
&= \quad (\frac{\partial f_3}{\partial y} - \frac{\partial f_2}{\partial z}, -(\frac{\partial f_3}{\partial x} - \frac{\partial f_1}{\partial z}), \frac{\partial f_2}{\partial x} - \frac{\partial f_1}{\partial y}).
\end{aligned}
$$

Note that

$$\text{curl} : (\text{Vector fields}) \to (\text{Vector fields}).$$

Now to look at an example and see that the curl is indeed measuring some sort of twirling. Earlier we saw that the vector field $\mathbf{F}(x, y, z) = (-y, x, 0)$ looks like a whirlpool. Its curl is:

$$
\text{curl}(\mathbf{F}) = \det \begin{pmatrix} i & j & k \\ \frac{\partial}{\partial x} & \frac{\partial}{\partial y} & \frac{\partial}{\partial z} \\ -y & x & 0 \end{pmatrix}.
$$

$$= (0, 0, 2),$$

which reflects that the whirlpool action is in the xy-plane, perpendicular to the z-axis.

We will see in the statement of Stokes' Theorem that intuitively the length of the curl($\mathbf{F}$) indeed measures how much the vector field is twirling about while the vector curl($\mathbf{F}$) points in the direction normal to the twirling.

5.1.8   Orientability

We also require our manifolds to be orientable. For a surface, orientability means that we can choose a normal vector field on the surface that varies continuously and never vanishes. For a curve, orientability means that we can choose a unit tangent vector, at each point, that varies continuously.

The standard example of a nonorientable surface is the Möbius strip, obtained by putting a half twist in a strip of paper and then attaching the ends.

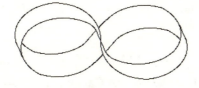

For an orientable manifold, there are always two choices of orientation, depending on which direction is chosen for the normal or the tangent. Further an oriented surface S with boundary curve ∂S will induce an orientation on ∂S, as will a 3-dimensional region induce an orientation on its boundary surface. If you happen to choose the wrong induced orientation for a boundary, the various versions of Stokes' Theorems will be off merely by a factor of (-1). Do not panic if you found the last few paragraphs vague. They were, deliberately so. To actually rigorously define orientation takes a little work. In first approaching the subject, it is best to concentrate on the basic examples and only then worry about the correct sign coming from the induced orientations. Rigorous definitions for orientability are given in the next chapter.

5.2 The Divergence Theorem and Stokes' Theorem

(For technical convenience, we will assume for the rest of this chapter that all functions, including those that make up vector fields, have as many derivatives as needed.)

The whole goal of this chapter is to emphasize that there must always be a deep link between the values of a function on the boundary of a manifold with the values of its derivative (suitably defined) on the interior of the manifold. This link is already present in

Theorem 5.2.1 (The Fundamental Theorem of Calculus) *Let*

$$f : [a, b] \to \mathbf{R}$$

be a a real-valued differentiable function on the interval $[a, b]$. Then

$$f(b) - f(a) = \int_a^b \frac{df}{dx}\,\mathrm{d}x.$$

Here the derivative $\frac{df}{dx}$ is integrated over the interval

$$[a, b] = \{x \in \mathbf{R} : a \le x \le b\},$$

which has as its boundary the points (a) and (b). The orientation on the boundary will be b and $-a$, or

$$\partial[a, b] = b - a.$$

Then the Fundamental Theorem of Calculus can be interpreted as stating that the value of $f(x)$ on the boundary is equal to the average (the integral) of the derivative over the interior.

One possible approach to generalizing the Fundamental Theorem is to replace the one-dimensional interval $[a, b]$ with something higher dimensional and replace the one variable function f with either a function of more than one variable or (less obviously) by a vector field. The correct generalizations will of course be determined by what can be proven.

In the divergence theorem, the interval becomes a three-dimensional manifold, whose boundary is a surface, and the function f becomes a vector field. The derivative of f will here be the divergence. More precisely:

Theorem 5.2.2 (The Divergence Theorem) *In* $\mathbf{R}^3$, *let M be a three-dimensional manifold with boundary ∂M a compact manifold of dimension two. Let $\mathbf{F}(x, y, z)$ denote a vector field on $\mathbf{R}^3$ and let $\mathbf{n}(x, y, z)$ denote a unit normal vector field to the boundary surface ∂M. Then*

$$\int\int_{\partial M} \mathbf{F} \cdot \mathbf{n} \, \mathrm{d}S = \int\int\int_M (div\mathbf{F}) \, \mathrm{d}x\mathrm{d}y\mathrm{d}z.$$

We will sketch a proof in section 5.5.

On the left hand side we have an integral of the vector field $\mathbf{F}$ over the boundary. On the right hand side we have an integral of the function $\mathrm{div}(\mathbf{F})$ (which involves derivatives of the vector field) over the interior.

In Stokes' Theorem, the interval becomes a surface, so that the boundary is a curve, and the function again becomes a vector field. The role of the derivative though will now be played by the curl of the vector field.

Theorem 5.2.3 (Stokes' Theorem) *Let M be a surface in $\mathbf{R}^3$ with compact boundary curve ∂M. Let $\mathbf{n}(x, y, z)$ be the unit normal vector field to M and let $\mathbf{T}(x, y, z)$ denote the induced unit tangent vector to the curve ∂M. If $\mathbf{F}(x, y, z)$ is any vector field, then*

$$\int_{\partial M} \mathbf{F} \cdot \mathbf{T} \, \mathrm{d}s = \int\int_M curl(\mathbf{F}) \cdot \mathbf{n} \, \mathrm{d}S.$$

As with the Divergence Theorem, a sketch of the proof will be given later in this chapter.

Again, on the left hand side we have an integral involving a vector field $\mathbf{F}$ on the boundary while on the right hand side we have an integral on the interior involving the curl of $\mathbf{F}$ (which is in terms of the various derivatives of $\mathbf{F}$).

Although both the Divergence Theorem and Stokes' Theorem were proven independently, their similarity is more than a mere analogy; both are special cases, as is the Fundamental Theorem of Calculus, of one very general theorem, which is the goal of the next chapter. The proofs of each are also quite similar. There are in fact two basic methods for proving these types of theorems. The first is to reduce to the Fundamental Theorem of Calculus, $f(b) - f(a) = \int_a^b \frac{df}{dx} dx$. This method will be illustrated in our sketch of the Divergence Theorem.

The second method involves two steps. Step one is to show that given two regions R_1 and R_2 that share a common boundary, we have

$$\int_{\partial R_1} \text{function} + \int_{\partial R_2} \text{function} = \int_{\partial(R_1 \cup R_2)} \text{function}.$$

Step two is to show that the theorem is true on infinitesimally small regions. To prove the actual theorem by this approach, simply divide the original region into infinitely many infinitesimally small regions, apply step two and then step one. We take this approach in our sketch of Stokes' Theorem.

Again, all of these theorems are really the same. In fact, to most mathematicians, these theorems usually go by the single name "Stokes' Theorem".

5.3 A Physical Interpretation of the Divergence Theorem

The goal of this section is to give a physical meaning to the Divergence Theorem, which was, in part, historically how the theorem was discovered. We will see that the Divergence Theorem states that the flux of a vector field through a surface is precisely equal to the sum of the divergences of each point of the interior. Of course, we need to give some definitions to these terms.

Definition 5.3.1 *Let S be a surface in $\mathbf{R}^3$ with unit normal vector field $\mathbf{n}(x, y, z)$. Then the flux of a vector field $\mathbf{F}(x, y, z)$ through the surface S is*

$$\int\int_S \mathbf{F} \cdot \mathbf{n} \, dS.$$

Intuitively we want the flux to measure how much of the vector field $\mathbf{F}$ pushes through the surface S.

Imagine a stream of water flowing along. The tangent vector of the direction of the water at each point defines a vector field $\mathbf{F}(x, y, z)$. Suppose the vector field $\mathbf{F}$ is:

Place into the stream an infinitely thin sheet of rubber, let us say. We want the flux to measure how hard it is to hold this sheet in place against the flow of the water. Here are three possibilities:

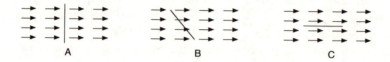

In case A, the water is hitting the rubber sheet head on, making it quite difficult to hold in place. In case C, no effort is needed to hold the sheet still, as the water just flows on by. The effort needed to keep the sheet still in case B is seen to be roughly halfway between effort needed in cases A and C. The key to somehow quantifying these differences of flux is to measure the angle between the vector field **F** of the stream and the normal vector field **n** to the membrane. Clearly, the dot product $\mathbf{F} \cdot \mathbf{n}$ works. Thus using that flux is defined by

$$\int\int_S \mathbf{F} \cdot \mathbf{n} \, dS,$$

the flux through surface A is greater than the flux through surface B which in turn is greater than the flux through surface C, which has flux equal to 0.

The Divergence Theorem states that the flux of a vector field through a boundary surface is exactly equal to the sum (integral) of the divergence of the vector field in the interior. In some sense the divergence must be an infinitesimal measure of the flux of a vector field.

5.4   A Physical Interpretation of Stokes' Theorem

Here we discuss the notion of the circulation of a vector field with respect to a curve. We will give the definition, then discuss what it means.

Definition 5.4.1 *Let C be a smooth curve in $\mathbf{R}^3$ with unit tangent vector field $\mathbf{T}(x, y, z)$. The circulation of a vector field $\mathbf{F}(x, y, z)$ along the curve C is*

$$\int_C \mathbf{F} \cdot \mathbf{T} \, ds.$$

Let **F** be a vector field representing a flowing stream of water, such as:

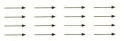

Put a thin wire (a curve C) into this stream with a small bead attached to it, with the bead free to move up and down the wire.

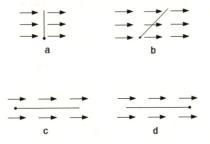

In case a, the water will not move the ball at all. In case b the ball will be pushed along the curve while in case c the water will move the ball the most quickly. In case d, not only will the ball not want to move along the curve C, effort is needed to even move the ball at all. These qualitative judgments are captured quantitatively in the above definition for circulation, since the dot product $\mathbf{F} \cdot \mathbf{T}$ measures at each point how much of the vector field $\mathbf{F}$ is pointing in the direction of the tangent $\mathbf{T}$ and hence how much of $\mathbf{F}$ is pointing in the direction of the curve.

In short, circulation measures how much of the vector field flows in the direction of the curve C. In physics, the vector field is frequently the force, in which case the circulation is a measurement of work.

Thus Stokes' Theorem is stating that the circulation of a vector field along a curve ∂M which bounds a surface M is precisely equal to the normal component of the vector field curl($\mathbf{F}$) in the interior. This is why the term 'curl' is used, as it measures the infinitesimal tendency of a vector field to have circulation, or in other words, it provides an infinitesimal measure of the "whirlpoolness" of the vector field.

5.5 Sketch of a Proof of the Divergence Theorem

This will only be a sketch, as we will be making a number of simplifying assumptions. First, assume that our three-dimensional manifold M (a solid) is *simple*, meaning that any line parallel to the $x-$axis, $y-$axis or $z-$axis can only intersect M in a connected line segment or a point. Thus

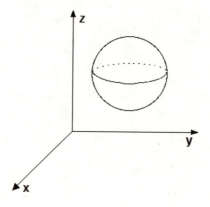

is simple while

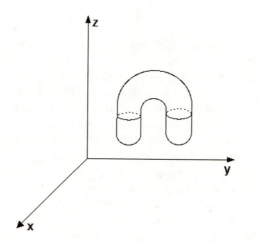

is not.

Denote the components of the vector field by

$$\mathbf{F}(z, y, z) = (f_1(x, y, z), f_2(x, y, z), f_3(x, y, z))$$
$$= (f_1, f_2, f_3).$$

On the boundary surface ∂M, denote the unit normal vector field by:

$$\mathbf{n}(x, y, z) = (n_1(x, y, z), n_2(x, y, z), n_3(x, y, z))$$
$$= (n_1, n_2, n_3).$$

We want to show that

$$\iint_{\partial M} \mathbf{F} \cdot \mathbf{n} \, dS = \iiint_M div(\mathbf{F}) dx dy dz.$$

In other words, we want

$$\iint_{\partial M} (f_1 n_1 + f_2 n_2 + f_3 n_3) dS = \iiint_M \left(\frac{\partial f_1}{\partial x} + \frac{\partial f_2}{\partial y} + \frac{\partial f_3}{\partial z} \right) dx dy dz.$$

If we can show

$$\iint_{\partial M} f_1 n_1 dS = \iiint_M \frac{\partial f_1}{\partial x} dx dy dz,$$

$$\iint_{\partial M} f_2 n_2 dS = \iiint_M \frac{\partial f_2}{\partial y} dx dy dz$$

$$\iint_{\partial M} f_3 n_3 dS = \iiint_M \frac{\partial f_3}{\partial z} dx dy dz$$

we will be done.

We will just sketch the proof of the last equation

$$\iint_{\partial M} f_3(x, y, z) n_3(x, y, z) dS = \iiint_M \frac{\partial f_3}{\partial z} dx dy dz,$$

since the other two equalities will hold for similar reasons.

The function $n_3(x, y, z)$ is the z-component of the normal vector field $\mathbf{n}(x, y, z)$. By the assumption that M is simple, we can split the boundary component ∂M into three connected pieces: $\{\partial M\}_{top}$, where $n_3 > 0$, $\{\partial M\}_{side}$, where $n_3 = 0$ and $\{\partial M\}_{bottom}$, where $n_3 < 0$.

For example, if ∂M is

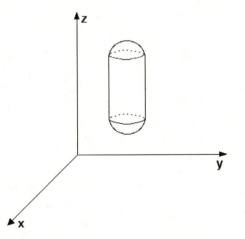

then

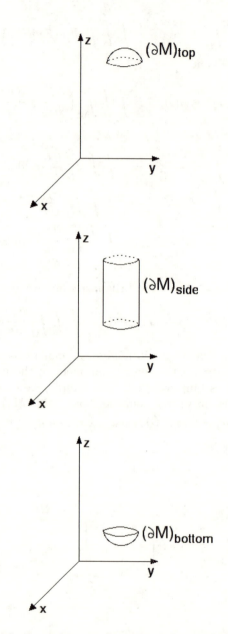

Then we can split the boundary surface integral into three parts:

$$\iint_{\partial M} f_3 n_3 \mathrm{d}S \;=\; \iint_{\partial M_{top}} f_3 n_3 \mathrm{d}S + \iint_{\partial M_{side}} f_3 n_3 \mathrm{d}S$$

$$+ \int\int_{\partial M_{bottom}} f_3 n_3 \mathrm{d}S$$

$$= \int\int_{\partial M_{top}} f_3 n_3 \mathrm{d}S + \int\int_{\partial M_{bottom}} f_3 n_3 \mathrm{d}S,$$

since n_3, the normal component in the z direction, will be zero on ∂M_{side}.

Further, again by the assumption of simplicity, there is a region R in the xy-plane such that $\{\partial M\}_{top}$ is the image of a function

$$(x, y) \rightarrow (x, y, t(x, y))$$

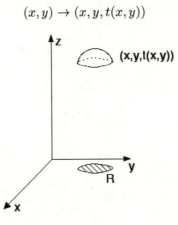

and $\{\partial M\}_{bottom}$ is the image of a function

$$(x, y) \rightarrow (x, y, b(x, y)).$$

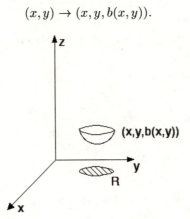

Then

$$\int\int_{\partial M} f_3 n_3 \mathrm{d}S = \int\int_{\partial M_{top}} f_3 n_3 \mathrm{d}S + \int\int_{\partial M_{bottom}} f_3 n_3 \mathrm{d}S$$

$$= \int\int_R f_3(x,y,t(x,y))\mathrm{d}x\mathrm{d}y +$$

$$\int\int_R f_3(x,y,b(x,y))\mathrm{d}x\mathrm{d}y$$

$$= \int\int_R (f_3(x,y,t(x,y)) - f_3(x,y,b(x,y)))\mathrm{d}x\mathrm{d}y,$$

where the minus sign in front of the last term coming from the fact that the normal to ∂M_{bottom} points downward. But this is just

$$\int\int_R \int_{b(x,y)}^{t(x,y)} \frac{\partial f_3}{\partial z}\mathrm{d}x\mathrm{d}y\mathrm{d}z,$$

by the Fundamental Theorem of Calculus. This, in turn, is equal to

$$\int\int\int_M \frac{\partial f_3}{\partial z}\mathrm{d}x\mathrm{d}y\mathrm{d}z,$$

which is what we wanted to show.

To prove the full result, we would need to take any solid M and show that we can split M into simple parts and then that if the Divergence Theorem is true on each simple part, it is true on the original M. While not intuitively difficult, this is nontrivial to prove and involves some subtle questions of convergence.

5.6   Sketch of a Proof for Stokes' Theorem

Let M be a surface with boundary curve ∂M.

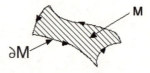

We break the proof of Stokes' Theorem into two steps. First, given two rectangles R_1 and R_2 that share a common side, we want

$$\int_{\partial R_1} \mathbf{F}\cdot\mathbf{T}\mathrm{d}s + \int_{\partial R_2} \mathbf{F}\cdot\mathbf{T}\mathrm{d}s = \int_{\partial R_1\cup R_2} \mathbf{F}\cdot\mathbf{T}\mathrm{d}s,$$

where $\mathbf{T}$ is the unit tangent vector.

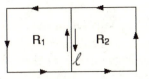

Second, we need to show that Stokes' Theorem is true on infinitesimally small rectangles.

The proof of the first is that for the common side ℓ of the two rectangles, the orientations are in opposite directions. This forces the value of the dot product $(\mathbf{F} \cdot \mathbf{T})$ along ℓ as a side of the rectangle R_1 to have opposite sign of the value of $(\mathbf{F} \cdot \mathbf{T})$ along ℓ as a side of the other rectangle R_2. Thus

$$\int_{\ell \subset \partial R_1} \mathbf{F} \cdot \mathbf{T} \, ds = - \int_{\ell \subset \partial R_2} \mathbf{F} \cdot \mathbf{T} \, ds.$$

Since the boundary of the union of the two rectangles $R_1 \cup R_2$ does not contain the side ℓ, we have

$$\int_{\partial R_1} \mathbf{F} \cdot \mathbf{T} \, ds + \int_{\partial R_2} \mathbf{F} \cdot \mathbf{T} \, ds = \int_{\partial R_1 \cup R_2} \mathbf{F} \cdot \mathbf{T} \, ds.$$

Before proving that Stokes' Theorem is true on infinitesimally small rectangles, assume for a moment that we already know this to be true. Split the surface M into (infinitely many) small rectangles.

Then

$$\int \int_M curl(\mathbf{F}) \cdot \mathbf{n} \, dS \quad = \quad \sum_{\text{small rectangles}} \int \int curl(\mathbf{F}) \cdot \mathbf{n} \, dS$$

$$= \sum \int_{\partial(\text{each rectangle})} \mathbf{F} \cdot \mathbf{T} \, ds,$$

since we are assuming that Stokes' Theorem is true on infinitesimally small rectangles. But by the first step, the above sum will equal to the single integral over the boundary of the union of the small rectangles

$$\int_{\partial M} \mathbf{F} \cdot \mathbf{T} \, ds,$$

which gives us Stokes' Theorem. Hence all we need to show is that Stokes' Theorem is true for infinitesimally small rectangles.

Before showing this, note that this argument is nonrigorous, as the whole sum is over infinitely many small rectangles, and thus subtle convergence questions would need to be solved. We pass over this in silence.

Now to sketch why Stokes' Theorem is true for infinitesimally small rectangles. This will also contain the justification for why the definition of the curl of a vector field is what it is.

By a change of coordinates, we can assume that our small rectangle R lies in the xy-plane with one vertex being the origin $(0,0)$.

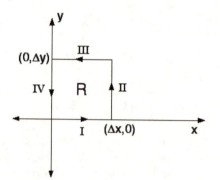

Its unit normal vector will be $\mathbf{n} = (0,0,1)$.

If the vector field is $\mathbf{F}(x,y,z) = (f_1, f_2, f_3)$, we have

$$curl(\mathbf{F}) \cdot \mathbf{n} = \frac{\partial f_2}{\partial x} - \frac{\partial f_1}{\partial y}.$$

We want to show that:

$$(\frac{\partial f_2}{\partial x} - \frac{\partial f_1}{\partial y})dxdy = \int_{\partial R} \mathbf{F} \cdot \mathbf{T}\, ds,$$

where $\mathbf{T}$ is the unit tangent vector to the boundary rectangle ∂R and $dx\, dy$ is the infinitesimal area for the rectangle R.

Now to calculate $\int_{\partial R} \mathbf{F} \cdot \mathbf{T}\, ds$.

The four sides of the rectangle ∂R have the following parametrizations.

Side	*Parametrization*	*Integral*
I :	$s(t) = (t\triangle x, 0), 0 \le t \le 1$	$\int_0^1 f_1(t\triangle x, 0)\triangle x dt$
II :	$s(t) = (\triangle x, t\triangle y), 0 \le t \le 1$	$\int_0^1 f_2(\triangle x, t\triangle y)\triangle y dt$
III :	$s(t) = (\triangle x - t\triangle x, \triangle y), 0 \le t \le 1$	$\int_0^1 -f_1(\triangle x - t\triangle x, \triangle y)\triangle x dt$
IV :	$s(t) = (0, \triangle y - t\triangle y), 0 \le t \le 1$	$\int_0^1 -f_2(0, \triangle y - t\triangle y)\triangle y dt$

It is always the case, for any function $f(t)$, that

$$\int_0^1 f(t)dt = \int_0^1 f(1-t)dt,$$

by changing the variable t to $1 - t$. Thus the integrals for sides III and IV can be replaced by $\int_0^1 -f_1(t\triangle x, \triangle y)\triangle x \; dt$ and $\int_0^1 -f_2(0, t\triangle y)\triangle y \; dt$. Then

$$
\begin{aligned}
\int_{\partial R} \mathbf{F} \cdot \mathbf{T} \, ds \;=\;& \int_I \mathbf{F} \cdot \mathbf{T} \, ds + \int_{II} \mathbf{F} \cdot \mathbf{T} \, ds + \int_{III} \mathbf{F} \cdot \mathbf{T} \, ds + \int_{IV} \mathbf{F} \cdot \mathbf{T} \, ds \\
=\;& \int_0^1 (f_1(t\triangle x, 0)\triangle x + f_2(\triangle x, t\triangle y)\triangle y \\
& - f_1(t\triangle x, \triangle y)\triangle x - f_2(0, t\triangle y)\triangle y)dt \\
=\;& \int_0^1 (f_2(\triangle x, t\triangle y) - f_2(0, t\triangle y))\triangle y dt \\
& - \int_0^1 (f_1(t\triangle x, \triangle y) - f_1(t\triangle x, 0))\triangle x dt \\
=\;& \int_0^1 \frac{f_2(\triangle x, t\triangle y) - f_2(0, t\triangle y)}{\triangle x} \\
& - \frac{f_1(t\triangle x, \triangle y) - f_1(t\triangle x, y)}{\triangle y})\triangle x\triangle y dt,
\end{aligned}
$$

which converges to

$$\int_0^1 (\frac{\partial f_2}{\partial x} - \frac{\partial f_1}{\partial y})d x d y dt,$$

as $\triangle x, \triangle y \to 0$. But this last integral will be

$$(\frac{\partial f_2}{\partial x} - \frac{\partial f_1}{\partial y})d x d y$$

which is what we wanted.

Again, letting $\triangle x, \triangle y \to 0$ is a nonrigorous step. Also, the whole nonchalant way in which we changed coordinates to put our rectangle into the xy-plane would have to be justified in a rigorous proof.

5.7   Books

Most calculus books have sections near the end on the multivariable calculus covered in this chapter. A long time popular choice is Thomas and Finney's text [36]. Another good source is Stewart's *Calculus* [108].

Questions in physics, especially in electricity and magnetism, were the main historical motivation for the development of the mathematics in this chapter. There are physical "proofs" of the Divergence Theorem and Stokes' Theorem. Good sources are in Halliday and Resnick's text in physics [51] and in Feynmann's *Lectures in Physics* [35].

5.8   Exercises

1. Extend the proof of the Divergence Theorem, given in this chapter for simple regions, to the region:

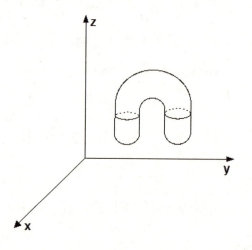

2. Let D be the disc of radius r, with boundary circle ∂D, given by the equations:
$$D = \{(x, y, 0) : x^2 + y^2 \le r\}.$$
For the vector field
$$\mathbf{F}(x, y, z) = (x + y + z, 3x + 2y + 4z, 5x - 3y + z),$$

find the path integral $\int_{\partial D} \mathbf{F} \cdot \mathbf{T}\ ds$, where $\mathbf{T}$ is the unit tangent vector of the circle ∂D.

3. Consider the vector field

$$\mathbf{F}(x, y, z) = (x, 2y, 5z).$$

Find the surface integral $\int \int_{\partial M} \mathbf{F} \cdot \mathbf{n}\ dS$, where the surface ∂M is the boundary of the ball

$$M = \{(x, y, z) : x^2 + y^2 + z^2 \leq r\}$$

of radius r centered at the origin and $\mathbf{n}$ is the unit normal vector.

4. Let S be the surface that is the image of the map

$$r : \mathbf{R}^2 \to \mathbf{R}^3$$

given by

$$r(u, v) = (x(u, v), y(u, v), z(u, v)).$$

Considering the image of the line $v = $ constant, justify to yourself that

$$(\frac{\partial x}{\partial u}, \frac{\partial y}{\partial u}, \frac{\partial z}{\partial u})$$

is a tangent vector to S.

5. Green's Theorem is:

Theorem 5.8.1 (Green's Theorem) *Let σ be a simple loop in* $\mathbf{C}$ *and* Ω *its interior. If $P(x, y)$ and $Q(x, y)$ are two real-valued differentiable functions, then*

$$\int_\sigma P\ dx + Q\ dy = \int \int_\Omega (\frac{\partial Q}{\partial x} - \frac{\partial P}{\partial y})\ dx\ dy.$$

By putting the region Ω into the plane $z = 0$ and letting our vector field be $\langle P(x, y), Q(x, y), 0 \rangle$, show that Green's Theorem follows from Stokes' Theorem.

Chapter 6

Differential Forms and Stokes' Theorem

Basic Objects:	Differential Forms and Manifolds
Basic Goal:	Stokes' Theorem

In the last chapter we saw various theorems, all of which related the values of a function on the boundary of a geometric object with the values of the function's derivative on the interior. The goal of this chapter is to show that there is a single theorem (Stokes' Theorem) underlying all of these results. Unfortunately, a lot of machinery is needed before we can even state this grand underlying theorem. Since we are talking about integrals and derivatives, we have to develop the techniques that will allow us to integrate on k-dimensional spaces. This will lead to differential forms, which are the objects on manifolds that can be integrated. The exterior derivative is the technique for differentiating these forms. Since integration is involved, we will have to talk about calculating volumes. This is done in section one. Section two defines differential forms. Section three links differential forms with the vector fields, gradients, curls and divergences from last chapter. Section four gives the definition of a manifold (actually, three different methods for defining manifolds are given). Section five concentrates on what it means for a manifold to be orientable. In section six, we define how to integrate a differential form along a manifold, allowing us finally in section seven to state and to sketch a proof of Stokes' Theorem.

6.1   Volumes of Parallelepipeds

In this chapter, we are ultimately interested in understanding integration on manifolds (which we have yet to define). This section, though, is pure linear algebra, but linear algebra that is crucial for the rest of the chapter.

The problem is the following: In $\mathbf{R}^n$, suppose we are given k vectors $\mathbf{v}_1, \ldots, \mathbf{v}_k$. These k vectors will define a parallelepiped in $\mathbf{R}^n$. The question is how to compute the volume of this parallelepiped. For example, consider the two vectors

$$\mathbf{v}_1 = \begin{pmatrix} 1 \\ 2 \\ 3 \end{pmatrix} \text{ and } \mathbf{v}_2 = \begin{pmatrix} 3 \\ 2 \\ 1 \end{pmatrix}.$$

The parallelepiped that these two vectors span is a parallelogram in $\mathbf{R}^3$. We want a formula to calculate the area of this parallelogram. (Note: the true three dimensional volume of this flat parallelogram is zero, in the same way that the length of a point is zero and that the area of a line is zero; we are here trying to measure the two-dimensional "volume" of this parallelogram.)

We already know the answer in two special cases. For a single vector

$$\mathbf{v} = \begin{pmatrix} a_1 \\ \vdots \\ a_n \end{pmatrix}$$

in $\mathbf{R}^n$, the parallelepiped is the single vector $\mathbf{v}$. Here by "volume" we mean the length of this vector, which is, by the Pythagorean theorem,

$$\sqrt{a_1^2 + \cdots + a_n^2}.$$

The other case is when we are given n vectors in $\mathbf{R}^n$. Suppose the n vectors are

$$\mathbf{v}_1 = \begin{pmatrix} a_{11} \\ \vdots \\ a_{n1} \end{pmatrix}, \ldots, \mathbf{v}_n = \begin{pmatrix} a_{1n} \\ \vdots \\ a_{nn} \end{pmatrix}.$$

Here we know that the volume of the resulting parallelepiped is

$$\left| \det \begin{pmatrix} a_{11} & \cdots & a_{1n} \\ & \vdots & \\ a_{n1} & \cdots & a_{nn} \end{pmatrix} \right|,$$

following from one of the definitions of the determinant given in Chapter One. Our eventual formula will yield both of these results.

We will first give the formula and then discuss why it is reasonable. Write the k vectors $\mathbf{v}_1, \ldots, \mathbf{v}_k$ as column vectors. Set

$$A = (\mathbf{v}_1, \ldots, \mathbf{v}_k),$$

a $k \times n$ matrix. We denote the transpose of A by A^T, the $n \times k$ matrix

$$A^T = \begin{pmatrix} \mathbf{v}_1^T \\ \vdots \\ \mathbf{v}_k^T \end{pmatrix}$$

where each $\mathbf{v}_i^T$ is the writing of the vector $\mathbf{v}_i$ as a row vector. Then

Theorem 6.1.1 *The volume of the parallelepiped spanned by the vectors* $\mathbf{v}_1, \ldots, \mathbf{v}_k$ *is*

$$\sqrt{\det(A^T A)}.$$

Before sketching a proof, let us look at some examples. Consider the single vector

$$\mathbf{v} = \begin{pmatrix} a_1 \\ \vdots \\ a_n \end{pmatrix}.$$

Here the matrix A is just $\mathbf{v}$ itself. Then

$$\sqrt{\det(A^T A)} = \sqrt{\det(\mathbf{v}^T \mathbf{v})}$$

$$= \sqrt{\det\left((a_1, \ldots, a_n) \begin{pmatrix} a_1 \\ \vdots \\ a_n \end{pmatrix}\right)}$$

$$= \sqrt{\det(a_1^2 + \cdots + a_n^2)}$$

$$= \sqrt{a_1^2 + \cdots + a_n^2},$$

the length of the vector $\mathbf{v}$.

Now consider the case of n vectors $\mathbf{v}_1, \ldots, \mathbf{v}_n$. Then the matrix A is $n \times n$. We will use that $\det(A) = \det(A^T)$. Then

$$\sqrt{\det(A^T A)} = \sqrt{\det(A^T) \det(A)}$$

$$= \sqrt{\det(A)^2}$$

$$= |\det(A)|,$$

as desired.

Now to see why in general $\sqrt{\det(A^T A)}$ must be the volume. We need a preliminary lemma that yields a more intrinsic, geometric approach to $\sqrt{\det(A^T A)}$.

Lemma 6.1.1 *For the matrix*

$$A = (\mathbf{v}_1, \ldots, \mathbf{v}_k),$$

we have that

$$A^T A = \begin{pmatrix} |\mathbf{v}_1|^2 & \mathbf{v}_1 \cdot \mathbf{v}_2 & \cdots & \mathbf{v}_1 \cdot \mathbf{v}_k \\ \vdots & \vdots & \vdots & \vdots \\ \mathbf{v}_k \cdot \mathbf{v}_1 & \mathbf{v}_k \cdot \mathbf{v}_2 & \cdots & |\mathbf{v}_k|^2 \end{pmatrix},$$

where $\mathbf{v}_i \cdot \mathbf{v}_j$ denotes the dot product of vectors $\mathbf{v}_i$ and $\mathbf{v}_j$ and $|\mathbf{v}_i| = \sqrt{\mathbf{v}_i \cdot \mathbf{v}_i}$ denotes the length of the vector $\mathbf{v}_i$.

The proof of this lemma is just looking at

$$A^T A = \begin{pmatrix} \mathbf{v}_1^T \\ \vdots \\ \mathbf{v}_k^T \end{pmatrix} (\mathbf{v}_1, \ldots, \mathbf{v}_k).$$

Notice that if we apply any linear transformation on $\mathbf{R}^n$ that preserves angles and lengths (in other words, if we apply a rotation to $\mathbf{R}^n$), the numbers $|\mathbf{v}_i|$ and $\mathbf{v}_i \cdot \mathbf{v}_j$ do not change. (The set of all linear transformations of $\mathbf{R}^n$ that preserve angles and lengths form a group that is called the *orthogonal* group and denoted by $O(n)$.) This will allow us to reduce the problem to the finding of the volume of a parallelepiped in $\mathbf{R}^k$.

Sketch of Proof of Theorem: We know that

$$\sqrt{\det(A^T A)} = \sqrt{\det \begin{pmatrix} |\mathbf{v}_1|^2 & \mathbf{v}_1 \cdot \mathbf{v}_2 & \cdots & \mathbf{v}_1 \cdot \mathbf{v}_k \\ \vdots & \vdots & \vdots & \vdots \\ \mathbf{v}_k \cdot \mathbf{v}_1 & \mathbf{v}_k \cdot \mathbf{v}_2 & \cdots & |\mathbf{v}_k|^2 \end{pmatrix}}.$$

We will show that this must be the volume. Recall the standard basis for $\mathbf{R}^n$:

$$\mathbf{e}_1 = \begin{pmatrix} 1 \\ 0 \\ \vdots \\ 0 \end{pmatrix}, \mathbf{e}_2 = \begin{pmatrix} 0 \\ 1 \\ \vdots \\ 0 \end{pmatrix}, \ldots, \mathbf{e}_n = \begin{pmatrix} 0 \\ 0 \\ \vdots \\ 1 \end{pmatrix}$$

We can find a rotation of $\mathbf{R}^n$ that both preserves lengths and angles and more importantly, rotates our vectors $\mathbf{v}_1, \ldots, \mathbf{v}_k$ so that they lie in the span of the first k standard vectors $\mathbf{e}_1, \ldots, \mathbf{e}_k$. (To rigorously show this takes some work, but it is geometrically reasonable.) After this rotation, the last $n - k$ entries for each vector $\mathbf{v}_i$ are zero. Thus we can view our parallelepiped as being formed from k vectors in $\mathbf{R}^k$. But we already know how to compute this; it is

$$\sqrt{\det \begin{pmatrix} |\mathbf{v}_1|^2 & \mathbf{v}_1 \cdot \mathbf{v}_2 & \cdots & \mathbf{v}_1 \cdot \mathbf{v}_k \\ \vdots & \vdots & \vdots & \vdots \\ \mathbf{v}_k \cdot \mathbf{v}_1 & \mathbf{v}_k \cdot \mathbf{v}_2 & \cdots & |\mathbf{v}_k|^2 \end{pmatrix}}.$$

We are done. □

6.2 Differential Forms and the Exterior Derivative

This will be a long and, at times, technical section. We will initially define elementary k-forms on $\mathbf{R}^n$, for which there is still clear geometric meaning. We will then use these elementary k-forms to generate general k-forms. Finally, and for now no doubt the most unintuitive part, we will give the definition for the exterior derivative, a device that will map k-forms to $(k + 1)$-forms and will eventually be seen to be a derivative-type operation. In the next section we will see that the gradient, the divergence and the curl of the last chapter can be interpreted in terms of the exterior derivative.

6.2.1 Elementary k-forms

We start with trying to understand elementary 2-forms in $\mathbf{R}^3$. Label the coordinate axis for $\mathbf{R}^3$ as x_1, x_2, x_3. There will be three elementary 2-forms, which will be denoted by $dx_1 \wedge dx_2$, $dx_1 \wedge dx_3$ and $dx_2 \wedge dx_3$. We must now determine what these symbols mean. (We will define 1-forms in a moment.)

In words, $dx_1 \wedge dx_2$ will measure the signed area of the projection onto the $x_1 x_2$-plane of any parallelepiped in $\mathbf{R}^3$, $dx_1 \wedge dx_3$ will measure the signed area of the projection onto the $x_1 x_3$-plane of any parallelepiped in $\mathbf{R}^3$ and $dx_2 \wedge dx_3$ will measure the signed area of the projection onto the $x_2 x_3$-plane of any parallelepiped in $\mathbf{R}^3$.

By looking at an example, we will see how to actually compute with these 2-forms. Consider two vectors in $\mathbf{R}^3$, labelled

$$\mathbf{v}_1 = \begin{pmatrix} 1 \\ 2 \\ 3 \end{pmatrix} \text{ and } \mathbf{v}_2 = \begin{pmatrix} 3 \\ 2 \\ 1 \end{pmatrix}.$$

These vectors span a parallelepiped P in $\mathbf{R}^3$. Consider the projection map $\pi : \mathbf{R}^3 \to \mathbf{R}^2$ of $\mathbf{R}^3$ to the $x_1 x_2$ plane. Thus

$$\pi(x_1, x_2, x_3) = (x_1, x_2).$$

We define $dx_1 \wedge dx_2$ acting on the parallelepiped P to be the area of $\pi(P)$. Note that

$$\pi(\mathbf{v}_1) = \begin{pmatrix} 1 \\ 2 \end{pmatrix} \quad \text{and} \quad \pi(\mathbf{v}_2) = \begin{pmatrix} 3 \\ 2 \end{pmatrix}.$$

Then $\pi(P)$ is the parallelogram:

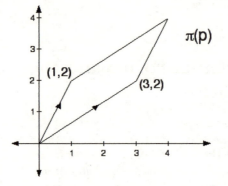

and the signed area is

$$
\begin{aligned}
dx_1 \wedge dx_2(P) &= \det(\pi(\mathbf{v}_1), \pi(\mathbf{v}_2)) \\
&= \det \begin{pmatrix} 1 & 3 \\ 2 & 2 \end{pmatrix} \\
&= -4
\end{aligned}
$$

In general, given a 3×2 matrix

$$A = \begin{pmatrix} a_{11} & a_{12} \\ a_{21} & a_{22} \\ a_{31} & a_{32} \end{pmatrix},$$

its two columns will define a parallelepiped. Then $dx_1 \wedge dx_2$ of this parallelepiped will be

$$dx_1 \wedge dx_2(A) = \det \begin{pmatrix} a_{11} & a_{12} \\ a_{21} & a_{22} \end{pmatrix}.$$

In the same way, $dx_1 \wedge dx_3$ will measure the area of the projection of a parallelepiped onto the $x_1 x_3$-plane. Then

$$dx_1 \wedge dx_3(A) = \det \begin{pmatrix} a_{11} & a_{12} \\ a_{31} & a_{32} \end{pmatrix}.$$

Likewise, we need

$$dx_2 \wedge dx_3(A) = \det \begin{pmatrix} a_{21} & a_{22} \\ a_{31} & a_{32} \end{pmatrix}.$$

Before defining elementary k-forms in general, let us look at elementary 1-forms. In $\mathbf{R}^3$, there are three elementary 1-forms, which will be denoted by dx_1, dx_2 and dx_3. Each will measure the one-dimensional volume (the length) of the projection of a one-dimensional parallelepiped in $\mathbf{R}^3$ to a coordinate axis. For example, with

$$\mathbf{v} = \begin{pmatrix} 1 \\ 2 \\ 3 \end{pmatrix},$$

its projection to the x_1-axis is just (1). Then we want to define

$$dx_1(\mathbf{v}) = dx_1 \begin{pmatrix} 1 \\ 2 \\ 3 \end{pmatrix} = 1.$$

In general, for a vector

$$\begin{pmatrix} a_{11} \\ a_{21} \\ a_{31} \end{pmatrix}$$

we have

$$dx_1 \begin{pmatrix} a_{11} \\ a_{21} \\ a_{31} \end{pmatrix} = a_{11}, \ dx_2 \begin{pmatrix} a_{11} \\ a_{21} \\ a_{31} \end{pmatrix} = a_{21}, \ dx_3 \begin{pmatrix} a_{11} \\ a_{21} \\ a_{31} \end{pmatrix} = a_{31}.$$

Now to define elementary k-forms on $\mathbf{R}^n$. Label the coordinates of $\mathbf{R}^n$ as $x_1, \ldots, x_n$. Choose an increasing subsequence of length k from $(1, 2, \ldots, n)$, which we will denote by

$$I = (i_1, \ldots, i_k)$$

with $1 \leq i_1 < \ldots < i_k \leq n$. Let

$$A = \begin{pmatrix} a_{11} & a_{12} & \cdots & a_{1k} \\ \vdots & \vdots & \vdots & \vdots \\ a_{n1} & \cdots & \cdots & a_{nk} \end{pmatrix}$$

be an $n \times k$ matrix. Its columns will span a k-dimensional parallelepiped P in $\mathbf{R}^n$. For convenience of exposition, let A_i be the ith row of A, i.e.,

$$A = \begin{pmatrix} A_1 \\ \vdots \\ A_n \end{pmatrix}.$$

We want the elementary k-form

$$\mathrm{d}x_I = \mathrm{d}x_{i_i} \wedge \cdots \wedge \mathrm{d}x_{i_k}$$

to act on the matrix A to give us the k-dimensional volume of the parallelepiped P projected onto the k-dimensional $x_{i_1}, \ldots, x_{i_k}$ space. This motivates the definition:

$$\mathrm{d}x_I(A) = \mathrm{d}x_{i_i} \wedge \cdots \wedge \mathrm{d}x_{i_k}(A) = \det \begin{pmatrix} A_{i_1} \\ \vdots \\ A_{i_k} \end{pmatrix}.$$

Elementary k-forms are precisely the devices that measure the volumes of k-dimensional parallelepipeds after projecting to coordinate k-spaces. The calculations come down to taking determinants of the original matrix with some of its rows deleted.

6.2.2   The Vector Space of k-forms

Recall back in Chapter One that we gave three different interpretations for the determinant of a matrix. The first was just how to compute it. The third was in terms of volumes of parallelepipeds, which is why determinants are showing up here. We now want to concentrate on the second interpretation, which in words was that the determinant is a multilinear map on the space of columns of a matrix. More precisely, if $M_{nk}(\mathbf{R})$ denotes the space of all $n \times k$ matrices with real entries, we had that the *determinant* of an $n \times n$ matrix A is defined as the unique real-valued function

$$\det : M_{nn}(\mathbf{R}) \to \mathbf{R}$$

satisfying:
  a) $\det(A_1, ..., \lambda A_k, ..., A_n) = \lambda \det(A_1, ..., A_k).$
  b) $\det(A_1, ..., A_k + \lambda A_i, ..., A_n) = \det(A_1, ..., A_n)$ for $k \neq i.$
  c) $\det(\text{Identity matrix}) = 1.$
A k-form will have a similar looking definition:

Definition 6.2.1 *A k-form ω is a real-valued function*

$$\omega : M_{nk}(\mathbf{R}) \to \mathbf{R}$$

satisfying:

$$\omega(A_1, ..., \lambda B + \mu C, ..., A_k) = \lambda \omega(A_1, ..., B, ..., A_k) + \mu \omega(A_1, ..., C, ..., A_k).$$

Thus ω is a multilinear real-valued function.

By the properties of determinants, we can see that each elementary k-form dx_I is in fact a k-form. (Of course this would have to be the case, or we wouldn't have called them elementary k-forms in the first place.) But in fact we have

Theorem 6.2.1 *The k-forms for a vector space $\mathbf{R}^n$ form a vector space of dimension $\binom{n}{k}$. The elementary k-forms are a basis for this vector space. This vector space is denoted by $\bigwedge^k(\mathbf{R}^n)$.*

We will not prove this theorem. It is not hard to prove that the k-forms are a vector space. It takes a bit more work to show that the elementary k-forms are a basis for $\bigwedge^k(\mathbf{R}^n)$.

Finally, note that 0-forms are just the real numbers themselves.

6.2.3   Rules for Manipulating k-forms

There is a whole machinery for manipulating k-forms. In particular, a k-form and an l-form can be combined to make a $(k+l)$-form. The method for doing this is not particularly easy to intuitively understand, but once you get the hang of it, it is a straightforward computational tool. We will look carefully at the $\mathbf{R}^2$ case, then describe the general rule for combining forms and finally see how this relates to the $\mathbf{R}^n$ case.

Let x_1 and x_2 be the coordinates for $\mathbf{R}^2$. Then dx_1 and dx_2 are the two elementary 1-forms and $dx_1 \wedge dx_2$ is the only elementary 2-form. But it looks, at least notationally, that the two 1-forms dx_1 and dx_2 somehow make up the 2-form $dx_1 \wedge dx_2$. We will see that this is indeed the case.

Let

$$\mathbf{v}_1 = \begin{pmatrix} a_{11} \\ a_{21} \end{pmatrix} \text{ and } \mathbf{v}_2 = \begin{pmatrix} a_{12} \\ a_{22} \end{pmatrix}$$

be two vectors in $\mathbf{R}^2$. Then

$$dx_1(\mathbf{v}_1) = a_{11} \text{ and } dx_1(\mathbf{v}_2) = a_{12}$$

and

$$dx_2(\mathbf{v}_1) = a_{21} \text{ and } dx_2(\mathbf{v}_2) = a_{22}.$$

The 2-form $dx_1 \wedge dx_2$ acting on the 2×2 matrix $(\mathbf{v}_1, \mathbf{v}_2)$ is the area of the parallelogram spanned by the vectors $\mathbf{v}_1$ and $\mathbf{v}_2$ and is hence the determinant of the matrix $(\mathbf{v}_1, \mathbf{v}_2)$. Thus

$$dx_1 \wedge dx_2(\mathbf{v}_1, \mathbf{v}_2) = a_{11}a_{22} - a_{12}a_{21}.$$

But note that this equals

$$dx_1(\mathbf{v}_1)\, dx_2(\mathbf{v}_2) - dx_1(\mathbf{v}_2)\, dx_2(\mathbf{v}_1).$$

At some level we have related our 2-form $dx_1 \wedge dx_2$ with our 1-forms dx_1 and dx_2, but it is not clear what is going on. In particular, at first glance it would seem to make more sense to change the above minus sign to a plus sign, but then, unfortunately, nothing would work out correctly.

We need to recall a few facts about the permutation group on n elements, S_n. (There is more discussion about permutations in Chapter Eleven.) Each element of S_n permutes the ordering of the set $\{1, 2, \ldots, n\}$. In general, every element of S_n can be expressed as the composition of flips (or transpositions).

If we need an even number of flips to express an element, we say that the element has sign 0 while if we need an odd number of flips, then the sign is 1. (Note that in order for this to be well-defined, we need to show that if an element has sign 0 (1), then it can only be written as the composition of an even (odd) number of flips; this is indeed true, but we will not show it.)

Consider S_2. There are only two ways we can permute the set $\{1, 2\}$. We can either just leave $\{1, 2\}$ alone (the identity permutation), which has sign 0, or flip $\{1, 2\}$ to $\{2, 1\}$, which has sign 1. We will denote the flip that sends $\{1, 2\}$ to $\{2, 1\}$ by $(1, 2)$. There are six ways of permuting the three elements $\{1, 2, 3\}$ and thus six elements in S_3. Each can be written as the composition of flips. For example, the permutation that sends $\{1, 2, 3\}$ to $\{3, 1, 2\}$ (which means that the first element is sent to the second slot, the second to the third slot and the third to the first slot) is the composition of the flip $(1, 2)$ with the flip $(1, 3)$, since, starting with $\{1, 2, 3\}$ and applying the flip $(1, 2)$, we get $\{2, 1, 3\}$. Then applying the flip $(1, 3)$ (which just interchanges the first and third elements), we get $\{3, 1, 2\}$.

We will use the following notational convention. If σ denotes the flip $(1, 2)$, then we say that

$$\sigma(1) = 2 \text{ and } \sigma(2) = 1.$$

Similarly, if σ denotes the element $(1, 2)$ composed with $(1, 3)$ in S_3, then we write

$$\sigma(1) = 2, \sigma(2) = 3 \text{ and } \sigma(3) = 1,$$

since under this permutation one is sent to two, two is sent to three and three is sent to one.

Suppose we have a k-form and an l-form. Let $n = k+l$. We will consider a special subset of S_n, the (k, l) shuffles, which are all elements $\sigma \in S_n$ that have the property that

$$\sigma(1) < \sigma(2) < \cdots < \sigma(k)$$

and

$$\sigma(k + 1) < \sigma(k + 2) < \cdots < \sigma(k + l).$$

Thus the element σ that is the composition of $(1,2)$ with $(1,3)$ is a $(2,1)$ shuffle, since

$$\sigma(1) = 2 < 3 = \sigma(2).$$

Denote the set of all (k,l) shuffles by $S(k,l)$. One of the exercises at the end of the chapter is to justify why these are called shuffles.

We can finally formally define the wedge product.

Definition 6.2.2 *Let* $A = (A_1, \ldots, A_{k+l})$ *be an* $N \times (k+l)$ *matrix, for any* N. *(Here each* A_i *denotes a column vector.) Let* τ *be a* k-*form and* ω *be an* l-*form. Then define*

$$\tau \wedge \omega(A) = \sum_{\sigma \in S(k,l)} (-1)^{sign(\sigma)} \tau(A_{\sigma(1)}, \ldots, A_{\sigma(k)}) \omega(A_{\sigma(k+1)}, \ldots, A_{\sigma(k+l)}).$$

Using this definition allows us to see that the wedge in $\mathbf{R}^2$ of two elementary 1-forms does indeed give us an elementary 2-form. A long calculation will show that in $\mathbf{R}^3$, the wedge of three elementary 1-forms yields the elementary 3-form.

It can be shown by these definitions that two 1-forms will anti-commute, meaning that

$$dx \wedge dy = -dy \wedge dx.$$

In general, we have that if τ is a k-form and ω is an l-form, then

$$\tau \wedge \omega = (-1)^{kl} \omega \wedge \tau.$$

This can be proven by directly calculating from the above definition of wedge product (though this method of proof is not all that enlightening). Note that for k and l both being odd, this means that

$$\tau \wedge \omega = (-1) \omega \wedge \tau.$$

Then for k being odd, we must have that

$$\tau \wedge \tau = (-1) \tau \wedge \tau,$$

which can only occur if

$$\tau \wedge \tau = 0.$$

In particular, this means that it is always the case that

$$dx_i \wedge dx_i = 0$$

and, if $i \neq j$,

$$dx_i \wedge dx_j = -dx_j \wedge dx_i.$$

6.2.4   Differential k-forms and the Exterior Derivative

Here the level of abstraction will remain high. We are after a general notion of what can be integrated (which will be the differential k-forms) and a general notion of what a derivative can be (which will be the exterior derivative).

First to define differential k-forms. In $\mathbf{R}^n$, if we let $I = \{i_1, \ldots, i_k\}$ denote some subsequence of integers with

$$1 \le i_1 < \ldots < i_k \le n,$$

then we let

$$\mathrm{d}x_I = \mathrm{d}x_{i_1} \wedge \cdots \wedge \mathrm{d}x_{i_k}.$$

Then a *differential k-form* ω is:

$$\omega = \sum_{\text{all possible } I} f_I \, \mathrm{d}x_I,$$

where each $f_I = f_I(x_1, \ldots, x_n)$ is a differentiable function.

Thus

$$(x_1 + \sin(x_2))\mathrm{d}x_1 + x_1 x_2 \mathrm{d}x_2$$

is an example of a differential 1-form, while

$$e^{x_1 + x_3} \, \mathrm{d}x_1 \wedge \mathrm{d}x_3 + x_2^3 \, \mathrm{d}x_2 \wedge \mathrm{d}x_3$$

is a differential 2-form.

Each differential k-form defines at each point of $\mathbf{R}^n$ a different k-form. For example, the differential 1-form $(x_1 + \sin(x_2))\mathrm{d}x_1 + x_1 x_2 \mathrm{d}x_2$ is the 1-form $3\,\mathrm{d}x_1$ at the point $(3, 0)$ and is $5\mathrm{d}x_1 + 2\pi\mathrm{d}x_2$ at the point $(4, \frac{\pi}{2})$.

To define the exterior derivative, we first define the exterior derivative of a differential 0-form and then by induction define the exterior derivative for a general differential k-form. We will see that the exterior derivative is a map from k-forms to $(k+1)$-forms:

$$\mathrm{d} : k\text{-forms} \ \rightarrow \ (k+1)\text{-forms}.$$

A differential 0-form is just another name for a differentiable function. Given a 0-form $f(x_1, \ldots, x_n)$, its *exterior derivative*, denoted by $\mathrm{d}f$, is:

$$\mathrm{d}f = \sum_{i=1}^n \frac{\partial f}{\partial x_i} \, \mathrm{d}x_i.$$

For example, if $f(x_1, x_2) = x_1 x_2 + x_2^3$, then

$$\mathrm{d}f = x_2 \mathrm{d}x_1 + (x_1 + 3x_2^2)\mathrm{d}x_2.$$

Note that the gradient of f is the similar looking $(x_2, x_1 + 3x_2^2)$. We will see in the next section that this is not chance.

Given a k-form $\omega = \sum_{\text{all possible } I} f_I \mathrm{d}x_I$, the *exterior derivative* $\mathrm{d}\omega$ is:

$$\mathrm{d}\omega = \sum_{\text{all possible } I} \mathrm{d}f_I \wedge \mathrm{d}x_I.$$

For example, in $\mathbf{R}^3$, let

$$\omega = f_1 \mathrm{d}x_1 + f_2 \mathrm{d}x_2 + f_3 \mathrm{d}x_3$$

be some 1-form. Then

$$
\begin{aligned}
\mathrm{d}\omega &= \mathrm{d}f_1 \mathrm{d}x_1 + \mathrm{d}f_2 \mathrm{d}x_2 + \mathrm{d}f_3 \mathrm{d}x_3 \\
&= (\frac{\partial f_1}{\partial x_1}\mathrm{d}x_1 + \frac{\partial f_1}{\partial x_2}\mathrm{d}x_2 + \frac{\partial f_1}{\partial x_3}\mathrm{d}x_3) \wedge \mathrm{d}x_1 \\
&\quad +(\frac{\partial f_2}{\partial x_1}\mathrm{d}x_1 + \frac{\partial f_2}{\partial x_2}\mathrm{d}x_2 + \frac{\partial f_2}{\partial x_3}\mathrm{d}x_3) \wedge \mathrm{d}x_2 \\
&\quad +(\frac{\partial f_3}{\partial x_1}\mathrm{d}x_1 + \frac{\partial f_3}{\partial x_2}\mathrm{d}x_2 + \frac{\partial f_3}{\partial x_3}\mathrm{d}x_3) \wedge \mathrm{d}x_3 \\
&= (\frac{\partial f_3}{\partial x_1} - \frac{\partial f_1}{\partial x_3})\mathrm{d}x_1 \wedge \mathrm{d}x_3 + (\frac{\partial f_2}{\partial x_1} - \frac{\partial f_1}{\partial x_2})\mathrm{d}x_1 \wedge \mathrm{d}x_2 \\
&\quad +(\frac{\partial f_3}{\partial x_2} - \frac{\partial f_2}{\partial x_3})\mathrm{d}x_2 \wedge \mathrm{d}x_3.
\end{aligned}
$$

Note that this looks similar to the curl of the vector field

$$(f_1, f_2, f_3).$$

Again, we will see that this similarity is not just chance.

Key to many calculations is:

Proposition 6.2.1 *For any differential k-form ω, we have*

$$\mathrm{d}(\mathrm{d}\omega) = 0.$$

The proof is one of the exercises at the end of the chapter, but you need to use that in $\mathbf{R}^n$ the order of differentiation does not matter, i.e.,

$$\frac{\partial}{\partial x_i}\frac{\partial f}{\partial x_j} = \frac{\partial}{\partial x_j}\frac{\partial f}{\partial x_i},$$

and that $\mathrm{d}x_i \wedge \mathrm{d}x_j = -\mathrm{d}x_j \wedge \mathrm{d}x_i$.

6.3   Differential Forms and Vector Fields

The overall goal for this chapter is to show that the classical Divergence Theorem, Green's Theorem and Stokes' Theorem are all special cases of one general theorem. This one general theorem will be stated in the language of differential forms. In order to see how it reduces to the theorems of last chapter, we need to relate differential forms with functions and vector fields. In $\mathbf{R}^3$, we will see that the exterior derivative, under suitable interpretation, will correspond to the gradient, the curl and the divergence.

Let x, y and z denote the standard coordinates for $\mathbf{R}^3$. Our first step is to define maps

$$
\begin{aligned}
T_0 &: \text{0-forms} &\rightarrow& \quad \text{functions on } \mathbf{R}^3 \\
T_1 &: \text{1-forms} &\rightarrow& \quad \text{vector fields on } \mathbf{R}^3 \\
T_2 &: \text{2-forms} &\rightarrow& \quad \text{vector fields on } \mathbf{R}^3 \\
T_3 &: \text{3-forms} &\rightarrow& \quad \text{functions on } \mathbf{R}^3.
\end{aligned}
$$

We will see that T_0, T_1 and T_3 have natural definitions. The definition for T_2 will take a bit of justification.

In the last section, we saw that differential 0-forms are just functions. Thus T_0 is just the identity map. From last section, we know that there are three elementary 1-forms: dx, dy and dz. Thus a general differential 1-form will be

$$\omega = f_1(x, y, z)dx + f_2(x, y, z)dy + f_3(x, y, z)dz,$$

where f_1, f_2 and f_3 are three separate functions on $\mathbf{R}^3$. Then define

$$T_1(\omega) = (f_1, f_2, f_3).$$

The definition for T_3 is just as straightforward. We know that on $\mathbf{R}^3$ there is only a single elementary 3-form, namely $dx \wedge dy \wedge dz$. Thus a general differential 3-form looks like:

$$\omega = f(x, y, z)dx \wedge dy \wedge dz,$$

where f is a function on $\mathbf{R}^3$. Then we let

$$T_3(\omega) = f(x, y, z).$$

As we mentioned, the definition for T_2 is not as straightforward. There are three elementary 2-forms: $dx \wedge dy$, $dx \wedge dz$ and $dy \wedge dz$. A general differential 2-form looks like:

$$\omega = f_1(x, y, z)dx \wedge dy + f_2(x, y, z)dx \wedge dz + f_3(x, y, z)dy \wedge dz,$$

where, as expected, f_1, f_2 and f_3 are functions on $\mathbf{R}^3$. Define the map T_2 by:

$$T_2(\omega) = (f_3, -f_2, f_1).$$

One method for justifying this definition will be that it will allow us to prove the theorems needed to link the exterior derivative with the gradient, the curl and the divergence. A second method will be in terms of dual spaces, as we will see in a moment.

We want to show:

Theorem 6.3.1 *On* $\mathbf{R}^3$, *let* ω_k *denote a* k-*form. Then*

$$T_1(\mathrm{d}\omega_0) = grad(T_0(\omega_0)),$$

$$T_2(\mathrm{d}\omega_1) = curl(T_1(\omega_1)),$$

and

$$T_3(\mathrm{d}\omega_2) = div(T_2(\omega_2)).$$

Each is a calculation (and is an exercise at the end of this chapter). We needed to define T_2 as we did in order to make the above work; this is one of the ways that we can justify our definition for the map T_2.

There is another justification for why T_2 must be what it is. This approach is a bit more abstract, but ultimately more important, as it generalizes to higher dimensions. Consider $\mathbf{R}^n$ with coordinates $x_1, \ldots, x_n$. There is only a single elementary n-form, namely $\mathrm{d}x_1 \wedge \ldots \wedge \mathrm{d}x_n$. Thus the vector space $\bigwedge^n(\mathbf{R}^n)$ of n-forms on $\mathbf{R}^n$ is one-dimensional and can be identified to the real numbers $\mathbf{R}$. Label this map by

$$T : \bigwedge\nolimits^n(\mathbf{R}^n) \to \mathbf{R}.$$

Thus $T(\alpha \mathrm{d}x_1 \wedge \ldots \wedge \mathrm{d}x_n) = \alpha$.

We now want to see that the dual vector space to $\bigwedge^k(\mathbf{R}^n)$ can be naturally identified with the vector space $\bigwedge^{n-k}(\mathbf{R}^n)$. Let ω_{n-k} be in $\bigwedge^{n-k}(\mathbf{R}^n)$. We first show how an $(n-k)$-form can be interpreted as a linear map on $\bigwedge^k(\mathbf{R}^n)$. If ω_k is any k-form, define

$$\omega_{n-k}(\omega_k) = T(\omega_{n-k} \wedge \omega_k).$$

It is a direct calculation that this is a linear map. From Chapter One we know that the dual vector space has the same dimension as the original vector space. By direct calculation, we also know that the dimensions for $\bigwedge^k(\mathbf{R}^n)$ and $\bigwedge^{n-k}(\mathbf{R}^n)$ are the same. Thus $\bigwedge^{n-k}(\mathbf{R}^n)$ is the dual space to $\bigwedge^k(\mathbf{R}^n)$.

Now consider the vector space $\bigwedge^1(\mathbf{R}^3)$, with its natural basis of $\mathrm{d}x, \mathrm{d}y$ and $\mathrm{d}z$. Its dual is then $\bigwedge^2(\mathbf{R}^3)$. As a dual vector space, an element of the natural basis is that which sends one of the basis vectors of $\bigwedge^1(\mathbf{R}^3)$ to one and the other basis vectors to zero. Thus the natural basis for $\bigwedge^2(\mathbf{R}^3)$, thought of as a dual vector space, is $dy \wedge dz$ (which corresponds to the 1-form dx, since $dy \wedge dz \wedge dx = 1 \cdot dx \wedge dy \wedge dz$), $-dx \wedge dz$ (which corresponds to dy) and $dx \wedge dy$ (which corresponds to dz). Then identifying dx with the row vector $(1,0,0)$, dy with $(0,1,0)$ and dz with $(0,0,1)$, we see that $dy \wedge dz$ should be identified with $(1,0,0)$, $dx \wedge dz$ with $(0,-1,0)$ and $dx \wedge dy$ with $(0,0,1)$. Then the 2-form

$$\omega = f_1 dx \wedge dy + f_2 dx \wedge dz + f_3 dy \wedge dz$$

should indeed be identified with $(f_3, -f_2, f_1)$, which is precisely how the map T_2 is defined.

6.4   Manifolds

While manifolds are to some extent some of the most natural occurring geometric objects, it takes work and care to create correct definitions. In essence, though, a k-dimensional manifold is any topological space that, in a neighborhood of any point, looks like a ball in $\mathbf{R}^k$. We will be at first concerned with manifolds that live in some ambient $\mathbf{R}^n$. For this type of manifold, we give two equivalent definitions: the parametric version and the implicit version. For each of these versions, we will carefully show that the unit circle S^1 in $\mathbf{R}^2$

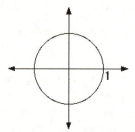

is a one-dimensional manifold. (Of course if we were just interested in circles we would not need all of these definitions; we are just using the circle to get a feel for the correctness of the definitions.) Then we will define an abstract manifold, a type of geometric object which need not be defined in terms of some ambient $\mathbf{R}^n$.

Consider again the circle S^1. Near any point $p \in S^1$ the circle looks like an interval (admittedly a bent interval). In a similar fashion, we want our definitions to yield that the unit sphere S^2 in $\mathbf{R}^3$ is a two-dimensional manifold, since near any point $p \in S^2$,

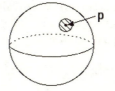

the sphere looks like a disc (though, again, more like a bent disc). We want to exclude from our definition of a manifold objects which contain points for which there is no well-defined notion of a tangent space, such as

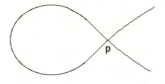

which has tangent difficulties at p, and the cone

which has tangent difficulties at the vertex p. As a technical note, we will throughout this section let M denote a second countable Hausdorff topological space.

For $k \leq n$, a k-dimensional *parametrizing* map is any differentiable map

$$\phi : (\text{Ball in } \mathbf{R}^k) \to \mathbf{R}^n$$

such that the rank of the Jacobian at every point is exactly k. In local coordinates, if $u_1, \ldots, u_k$ are the coordinates for $\mathbf{R}^k$ and if ϕ is described by the n differentiable functions $\phi_1, \ldots, \phi_n$ (i.e., $\phi = (\phi_1, \ldots, \phi_n)$), we require that at all points there is a $k \times k$ minor of the $n \times k$ Jacobian matrix

$$D\phi = \begin{pmatrix} \frac{\partial \phi_1}{\partial u_1} & \cdots & \frac{\partial \phi_1}{\partial u_k} \\ \vdots & & \vdots \\ \frac{\partial \phi_n}{\partial u_1} & \cdots & \frac{\partial \phi_n}{\partial u_k} \end{pmatrix}$$

that is invertible.

Definition 6.4.1 (Parametrized Manifolds) *The Hausdorff topological space M in $\mathbf{R}^n$ is a k-dimensional manifold if for every point $p \in M$ in $\mathbf{R}^n$, there is an open set U in $\mathbf{R}^n$ containing the point p and a parametrizing map ϕ such that*

$$\phi(\text{Ball in } \mathbf{R}^k) = M \cap U.$$

Consider the circle S^1. At the point $p = (1, 0)$, a parametrizing map is:

$$\phi(u) = (\sqrt{1 - u^2}, u),$$

while for the point $(0, 1)$, a parametrizing map could be:

$$\phi(u) = (u, \sqrt{1 - u^2}).$$

Given the parametrization, we will see in section five that it is easy to find a basis for the tangent space of the manifold. More precisely the tangent space is spanned by the columns of the Jacobian $D\phi$. This is indeed one of the computational strengths of using parametrizations for defining a manifold.

Another approach is to define a manifold as the zero locus of a set of functions on $\mathbf{R}^n$. Here the normal vectors are practically given to us in the definition.

Definition 6.4.2 (Implicit Manifolds) *A set M in $\mathbf{R}^n$ is a k-dimensional manifold if, for any point $p \in M$ there is an open set U containing p and $(n - k)$ differentiable functions $\rho_1, \ldots, \rho_{n-k}$ such that*

1. *$M \cap U = (\rho_1 = 0) \cap \cdots \cap (\rho_{n-k} = 0)$.*

2. *At all points in $M \cap U$, the gradient vectors*

$$\nabla \rho_1, \ldots, \nabla \rho_{n-k}$$

   are linearly independent.

It can be shown that the normal vectors are just the various $\nabla \rho_j$.

For an example, turn again to the circle S^1. The implicit method just notes that

$$S^1 = \{(x, y) : x^2 + y^2 - 1 = 0\}.$$

Here we have $\rho = x^2 + y^2 - 1$. Since

$$\nabla(x^2 + y^2 - 1) = (2x, 2y)$$

is never the zero vector, we are done.

The two definitions are equivalent, as discussed in the section on the implicit function theorem. But both of these definitions depend on our set M being in $\mathbf{R}^n$. Both critically use the properties of this ambient $\mathbf{R}^n$. There are situations where we still want to do calculus on a set of points which do not seem to live, in any natural way, in some $\mathbf{R}^n$. Historically this was first highlighted in Einstein's General Theory of Relativity, in which the universe itself was described as a 4-dimensional manifold that is neither $\mathbf{R}^4$ nor living in any natural way in a higher dimensional $\mathbf{R}^n$. By all accounts, Einstein was amazed that mathematicians had built up the whole needed machinery. Our goal here is to give the definition of an abstract manifold and then to show, once again, that S^1 is a manifold. Throughout this we will be using that we already know what it means for a function $f : \mathbf{R}^n \to \mathbf{R}^n$ to be differentiable.

Definition 6.4.3 (Manifolds) *A second countable Hausdorff topological space M is an n-dimensional manifold if there is an open cover (U_α) such that for each open set, U_α, we have a continuous map*

$$\phi_\alpha : Open \ ball \ in \ \mathbf{R}^n \to U_\alpha$$

that is one-to-one and onto and such that the map

$$\phi_\alpha^{-1}\phi_\beta : \phi_\beta^{-1}(U_\alpha \cap U_\beta) \to \phi_\alpha^{-1}(U_\alpha \cap U_\beta)$$

is differentiable.

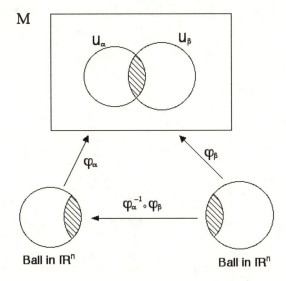

Note that $\phi_\beta^{-1}(U_\alpha \cap U_\beta)$ and $\phi_\alpha^{-1}(U_\alpha \cap U_\beta)$ are both open sets in $\mathbf{R}^n$ and thus we do know what it means for $\phi_\alpha^{-1}\phi_\beta$ to be differentiable, as discussed in Chapter Three. The idea is that we want to identify each open set U_α in M with its corresponding open ball in $\mathbf{R}^n$. In fact, if $x_1, \ldots, x_n$ are coordinates for $\mathbf{R}^n$, we can label every point p in U_α as the n-tuple given by $\phi_\alpha^{-1}(p)$. Usually people just say that we have chosen a coordinate system for U_α and identify it with the coordinates $x_1, \ldots, x_n$ for $\mathbf{R}^n$. It is this definition that motivates mathematicians to say that a manifold is anything that locally, around each point, looks like an open ball in $\mathbf{R}^n$.

Let us now show that S^1 satisfies this definition of a manifold. We will find an open cover of S^1 consisting of four open sets, for each of these write down the corresponding map ϕ_i and then see that $\phi_1^{-1}\phi_2$ is differentiable. (It is similar to show that the other $\phi_i^{-1}\phi_j$ are differentiable.)

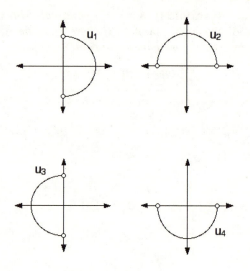

Set
$$U_1 = \{(x,y) \in S^1 : x > 0\}$$
and let
$$\phi_1 : (-1,1) \to U_1$$
be defined by
$$\phi_1(u) = (\sqrt{1-u^2}, u).$$
Here $(-1,1)$ denotes the open interval $\{x : -1 < x < 1\}$. In a similar fashion, set
$$U_2 = \{(x,y) \in S^1 : y > 0\}$$

$$U_3 = \{(x,y) \in S^1 : x < 0\}$$
$$U_4 = \{(x,y) \in S^1 : y < 0\}$$

and

$$\phi_2(u) = (u, \sqrt{1-u^2})$$
$$\phi_3(u) = (-\sqrt{1-u^2}, u)$$
$$\phi_4(u) = (u, -\sqrt{1-u^2}).$$

Now to show on the appropriate domain that $\phi_1^{-1}\phi_2$ is differentiable. We have

$$\phi_1^{-1}\phi_2(u) = \phi_1^{-1}(u, \sqrt{1-u^2}) = \sqrt{1-u^2}$$

which is indeed differentiable for $-1 < u < 1$. (The other verifications are just as straightforward.)

We can now talk about what it means for a function to be differentiable on a manifold. Again, we will reduce the definition to a statement about the differentiability of a function from $\mathbf{R}^n$ to $\mathbf{R}$.

Definition 6.4.4 *A real-valued function f on a manifold M is differentiable if for an open cover (U_α) and maps ϕ_α : Open ball in $\mathbf{R}^n \to U_\alpha$, the composition function*

$$f \circ \phi_\alpha : \text{Open ball in } \mathbf{R}^n \to \mathbf{R}$$

is differentiable.

There is still one difficulty with our abstract definition of a manifold. The definition depends upon the existence of an open cover of M. Think of our open cover of the circle S^1. Certainly there are many other open covers that will also place a manifold structure on S^1, such as:

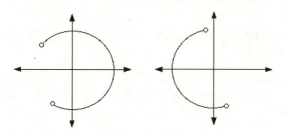

but still, it's the same circle. How can we identify these different ways of putting a manifold structure on the circle? We are led to the desire to find a natural notion of equivalence between manifolds (as we will see, we will denote this type of equivalence by saying that two manifolds are *diffeomorphic*). Before giving a definition, we need to define what it means to have a differentiable map between two manifolds. For notation, let M be an m-dimensional manifold with open cover (U_α) and corresponding maps ϕ_α and let N be an n-dimensional manifold with open cover (V_β) and corresponding maps η_β.

Definition 6.4.5 *Let $f : M \to N$ be a map from M to N. Let $p \in M$ with U_α an open set containing p. Set $q = f(p)$ and suppose that V_β is an open set containing q. Then f is differentiable at p if the map $\eta_\beta^{-1} \circ f \circ \phi_\alpha$ is differentiable in a neighborhood of the point $\phi_\alpha^{-1}(p)$ in $\mathbf{R}^m$. The map f is differentiable if it is differentiable at all points.*

We can now define our notion of equivalence.

Definition 6.4.6 *Two manifolds M and N are diffeomorphic if there exists a map $f : M \to N$ that is one-to-one, onto, differentiable and such that the inverse map, f^{-1}, is differentiable.*

Finally, by replacing the requirement that the various functions involved are differentiable by continuous functions, analytic functions, etc., we can define *continuous* manifolds, *analytic* manifolds, etc.

6.5   Tangent Spaces and Orientations

Before showing how to integrate differential k-forms along a k-dimensional manifold, we have to tackle the entirely messy issue of orientability. But before we can define orientability, we must define the tangent space to a manifold. If we use the implicit or parametric definition for a manifold, this will be straightforward. The definition for an abstract manifold is quite a bit more complicated (but as with most good abstractions, it is ultimately the right way to think about tangent vectors).

6.5.1   Tangent Spaces for Implicit and Parametric Manifolds

Let M be an implicitly defined manifold in $\mathbf{R}^n$ of dimension k. Then by definition, for each point $p \in M$ there is an open set U containing p and $(n-k)$ real-valued functions $\rho_1, \ldots, \rho_{n-k}$ defined on U such that

$$(\rho_1 = 0) \cap \ldots \cap (\rho_{n-k} = 0) = M \cap U$$

and, at every point $q \in M \cap U$, the vectors

$$\nabla \rho_1(q), \ldots, \nabla \rho_{n-k}(q)$$

are linearly independent. We have

Definition 6.5.1 *The normal space $N_p(M)$ to M at the point p is the vector space spanned by the vectors*

$$\nabla \rho_1(p), \dots, \nabla \rho_{n-k}(p).$$

The tangent space $T_p(M)$ to the manifold M at the point p consists of all vectors $\mathbf{v}$ in $\mathbf{R}^n$ that are perpendicular to each of the normal vectors.

If $x_1, \dots, x_n$ are the standard coordinates for $\mathbf{R}^n$, we have

Lemma 6.5.1 *A vector $\mathbf{v} = (v_1, \dots, v_n)$ is in the tangent space $T_p(M)$ if for all $i = 1, \dots, n - k$ we have*

$$0 = \mathbf{v} \cdot \nabla \rho_i(p) = \sum_{j=1}^{n} \frac{\partial \rho_i(p)}{\partial x_j} v_j.$$

The definition for the tangent space for parametrically defined manifolds is as straightforward. Here the Jacobian of the parametrizing map will be key. Let M be a manifold in $\mathbf{R}^n$, with the parametrizing map

$$\phi : (\text{Ball in } \mathbf{R}^k) \to \mathbf{R}^n$$

given by the n functions

$$\phi = (\phi_1, \dots, \phi_n).$$

The Jacobian for ϕ is the $n \times k$ matrix

$$D\phi = \begin{pmatrix} \frac{\partial \phi_1}{\partial u_1} & \cdots & \frac{\partial \phi_1}{\partial u_k} \\ \vdots & & \vdots \\ \frac{\partial \phi_n}{\partial u_1} & \cdots & \frac{\partial \phi_n}{\partial u_k} \end{pmatrix}.$$

Definition 6.5.2 *The tangent space $T_p(M)$ for M at the point p is spanned by the columns of the matrix $D\phi$.*

The equivalence of these two approaches can, of course, be shown.

6.5.2 Tangent Spaces for Abstract Manifolds

Both implicitly and parametrically defined manifolds live in an ambient $\mathbf{R}^n$, which carries with it a natural vector space structure. In particular, there is a natural notion for vectors in $\mathbf{R}^n$ to be perpendicular. We used this ambient space to define tangent spaces. Unfortunately, no such ambient

$\mathbf{R}^n$ exists for an abstract manifold. What we do know is what it means for a real-valued function to be differentiable.

In calculus, we learn about differentiation as a tool to both find tangent lines and also to compute rates of change of functions. Here we concentrate on the derivative as a rate of change. Consider three-space, $\mathbf{R}^3$, with the three partial derivatives $\frac{\partial}{\partial x}, \frac{\partial}{\partial y}$ and $\frac{\partial}{\partial z}$. Each corresponds to a tangent direction for $\mathbf{R}^3$ but each also gives a method for measuring how fast a function $f(x, y, z)$ is changing, i.e.,

$$\frac{\partial f}{\partial x} = \text{how fast f is changing in the x-direction,}$$

$$\frac{\partial f}{\partial y} = \text{how fast f is changing in the y-direction}$$

and

$$\frac{\partial f}{\partial z} = \text{how fast f is changing in the z-direction.}$$

This is how we are going to define tangent vectors on an abstract manifold, as rates of change for functions. We will abstract out the algebraic properties of derivatives (namely that they are linear and satisfy Leibniz's rule).

But we have to look at differentiable functions on M a bit more closely. If we want to take the derivative of a function f at a point p, we want this to measure the rate of change of f at p. This should only involve the values of f near p. What values f achieves away from p should be irrelevant. This is the motivation behind the following equivalence relation. Let (f, U) denote an open set on M containing p and a differentiable function f defined on U. We will say that

$$(f, U) \sim (g, V)$$

if, on the open set $U \cap V$, we have $f = g$. This leads us to defining

$$C_p^\infty = \{(f, U)\}/ \sim .$$

We will frequently abuse notation and denote an element of C_p^∞ by f. The space C_p^∞ is a vector space and captures the properties of functions close to the point p. (For mathematical culture sake, C_p^∞ is an example of a germ of a sheaf, in this case the sheaf of differentiable functions.)

Definition 6.5.3 *The tangent space $T_p(M)$ is the space of all linear maps*

$$v : C_p^\infty \to C_p^\infty$$

such that

$$v(fg) = fv(g) + gv(f).$$

To finish the story, we would need to show that this definition agrees with the other two, but this we leave as nontrivial exercises.

6.5.3 Orientation of a Vector Space

Our goal is to see that there are two possible orientations for any given vector space V. Our method is to set up an equivalence relation on the possible bases for V and see that there are only two equivalence classes, each of which we will call an orientation.

Let $\mathbf{v}_1, \ldots, \mathbf{v}_n$ and $\mathbf{w}_1, \ldots, \mathbf{w}_n$ be two bases for V. Then there exists unique real numbers a_{ij}, with $i, j = 1, \ldots, n$ such that

$$\mathbf{w}_1 = a_{11}\mathbf{v}_1 + \cdots + a_{1n}\mathbf{v}_n$$
$$\vdots$$
$$\mathbf{w}_n = a_{n1}\mathbf{v}_1 + \cdots + a_{nn}\mathbf{v}_n.$$

Label the $n \times n$ matrix (a_{ij}) by A. Then we know that $\det(A) \neq 0$. We say that the bases $\mathbf{v}_1, \ldots, \mathbf{v}_n$ and $\mathbf{w}_1, \ldots, \mathbf{w}_n$ have the same orientation if $\det(A) > 0$. If $\det(A) < 0$, then we say that they two bases have opposite orientation. It can be shown via matrix multiplication that

Lemma 6.5.2 *Having the same orientation is an equivalence relation on the set of bases for a vector space.*

The intuition is that two bases $\mathbf{v}_1, \ldots, \mathbf{v}_n$ and $\mathbf{w}_1, \ldots, \mathbf{w}_n$ should have the same orientation if we can continuously move the basis $\mathbf{v}_1, \ldots, \mathbf{v}_n$ to $\mathbf{w}_1, \ldots, \mathbf{w}_n$ so that at each step we still have a basis. In pictures for $\mathbf{R}^2$, the bases $\{(1,0), (0,1)\}$ and $\{(1,1), (-1,1)\}$ have the same orientation but different from the basis $\{(-1,0), (0,1)\}$.

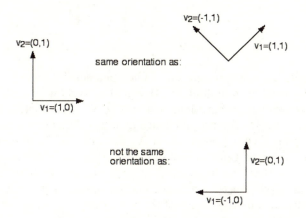

Choosing an orientation for a vector space means choosing one of the two possible orientations, i.e., choosing some basis.

6.5.4   Orientation of a Manifold and its Boundary

A manifold M has an orientation if we can choose a smoothly varying orientation for each tangent space $T_p(M)$. We ignore the technicalities of what 'smoothly varying' means, but the idea is that we can move our basis in a smooth manner from point to point on the manifold M.

Now let X^o be an open connected set in our oriented manifold M such that if X denotes the closure of X^o, then the boundary $\partial(X) = X - X^o$ is a smooth manifold of dimension one less than M. For example, if $M = \mathbf{R}^2$, an example of an X^o could be the open unit disc

$$D = \{(x,y) : x^2 + y^2 < 1\}.$$

Then the boundary of D is the unit circle

$$S^1 = \{(x,y) : x^2 + y^2 = 1\},$$

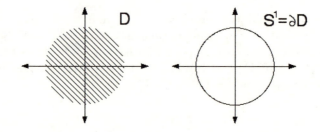

which is a one-dimensional manifold. The open set X^o inherits an orientation from the ambient manifold M. Our goal is to show that the boundary $\partial(X)$ has a canonical orientation. Let $p \in \partial(X)$. Since $\partial(X)$ has dimension one less than M, the normal space at p has dimension one. Choose a normal direction n that points out of X, not into X. The vector n, while normal to $\partial(X)$, is a tangent vector to M. Choose a basis $v_1, \ldots, v_{n-1}$ for $T_p(\partial(X))$ so that the basis $n, v_1, \ldots, v_{n-1}$ agrees with the orientation of M. It can be shown that all such chosen bases for $T_p(\partial(X))$ have the same orientation; thus the choice of the vectors $v_1, \ldots, v_{n-1}$ determines an orientation on the boundary manifold $\partial(X)$.

For example, let $M = \mathbf{R}^2$. At each point of $\mathbf{R}^2$, choose the basis $\{(1,0), (0,1)\}$.

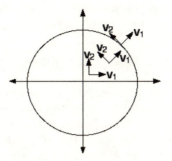

For the unit circle S^1, an outward pointing normal is always, at each point $p = (x, y)$, just the vector (x, y). Then the tangent vector $(-y, x)$ will give us a basis for $\mathbf{R}^2$ that has the same orientation as the given one. Thus we have a natural choice of orientation for the boundary manifold.

6.6 Integration on Manifolds

The goal of this section is to make sense out of the symbol

$$\int_M \omega,$$

where M will be a k-dimensional manifold and ω will be a differential k-form. Thus we want to (finally) show that differential k-forms are the things that will integrate along k-dimensional manifolds. The method will be to reduce all calculations to doing multiple integrals on $\mathbf{R}^k$, which we know how to do.

We will first look carefully at the case of 1-forms on $\mathbf{R}^2$. Our manifolds will be 1-dimensional and hence curves. Let C be a curve in the plane $\mathbf{R}^2$ that is parametrized by the map:

$$\sigma : [a, b] \to \mathbf{R}^2,$$

with

$$\sigma(u) = (x(u), y(u)).$$

If $f(x, y)$ is a continuous function defined on $\mathbf{R}^2$, then define the path integral, $\int_C f(x, y)\mathrm{d}x$, by the formula

$$\int_C f(x, y)\mathrm{d}x = \int_a^b f(x(u), y(u))\frac{\mathrm{d}x}{\mathrm{d}u}\mathrm{d}u.$$

Note that the second integral is just a one-variable integral over an interval on the real line. Likewise, the symbol $\int_C f(x,y)\mathrm{d}y$ is interpreted as

$$\int_C f(x,y)\mathrm{d}y = \int_a^b f(x(u),y(u))\frac{\mathrm{d}y}{\mathrm{d}u}\mathrm{d}u.$$

Using the chain rule, it can be checked that the numbers $\int_C f(x,y)\mathrm{d}x$ and $\int_C f(x,y)\mathrm{d}y$ are independent of the chosen parametrizations. Both of these are highly suggestive, as at least formally $f(x,y)\mathrm{d}x$ and $f(x,y)\mathrm{d}y$ look like differential 1-forms on the plane $\mathbf{R}^2$. Consider the Jacobian of the parametrizing map $\sigma(u)$, which is the 2×1 matrix

$$D\sigma = \begin{pmatrix} \mathrm{d}x/\mathrm{d}u \\ \mathrm{d}y/\mathrm{d}u \end{pmatrix}.$$

Letting $f(x,y)\mathrm{d}x$ and $f(x,y)\mathrm{d}y$ be differential 1-forms, we have by definition that at each point of $\sigma(u)$,

$$f(x,y)\mathrm{d}x(D\sigma) = f(x,y)\mathrm{d}x\left(\begin{pmatrix} \mathrm{d}x/\mathrm{d}u \\ \mathrm{d}y/\mathrm{d}u \end{pmatrix}\right) = f(x(u),y(u))\frac{\mathrm{d}x}{\mathrm{d}u}$$

and

$$f(x,y)\mathrm{d}y(D\sigma) = f(x,y)\mathrm{d}y\left(\begin{pmatrix} \mathrm{d}x/\mathrm{d}u \\ \mathrm{d}y/\mathrm{d}u \end{pmatrix}\right) = f(x(u),y(u))\frac{\mathrm{d}y}{\mathrm{d}u}.$$

Thus we could write the integrals $\int_C f(x,y)\mathrm{d}x$ and $\int_C f(x,y)\mathrm{d}y$ as

$$\int_C f(x,y)\mathrm{d}x = \int_a^b f(x,y)\mathrm{d}x(D\sigma)\mathrm{d}u$$

and

$$\int_C f(x,y)\mathrm{d}y = \int_a^b f(x,y)\mathrm{d}y(D\sigma)\mathrm{d}u.$$

This suggests how to define in general $\int_M \omega$. We will use that ω, as a k-form, will send any $n \times k$ matrix to a real number. We will parametrize our manifold M and take ω of the Jacobian of the parametrizing map.

Definition 6.6.1 *Let M be a k-dimensional oriented differentiable manifold in $\mathbf{R}^n$ such that there is a parametrizing one-to-one onto map*

$$\phi : B \to M$$

where B denotes the unit ball in $\mathbf{R}^k$. Suppose further that the parametrizing map agrees with the orientation of the manifold M. Let ω be a differential k-form on $\mathbf{R}^n$. Then

$$\int_M \omega = \int_B \omega(D\phi)\mathrm{d}u_1\cdots\mathrm{d}u_k.$$

Via a chain rule calculation, we can show that $\int_M \omega$ is well-defined:

Lemma 6.6.1 *Given two orientation preserving parametrizations ϕ_1 and ϕ_1 of a k-dimensional manifold M, we have*

$$\int_B \omega(D\phi_1)\mathrm{d}u_1 \cdots \mathrm{d}u_k = \int_B \omega(D\phi_2)\mathrm{d}u_1 \cdots \mathrm{d}u_k.$$

Thus $\int_M \omega$ is independent of parametrization.

We now know what $\int_M \omega$ means for a manifold that is the image of a differentiable one-to-one onto map from a ball in $\mathbf{R}^k$. Not all manifolds can be written as the image of a single parametrizing map. For example, the unit sphere S^2 in $\mathbf{R}^3$ needs at least two such maps (basically to cover both the north and south poles). But we can (almost) cover reasonable oriented manifolds by a countable collection of non-overlapping parametrizations. More precisely, we can find a collection $\{U_\alpha\}$ of nonoverlapping open sets in M such that for each α there exists a parametrizing orientation preserving map

$$\phi_\alpha : B \to U_\alpha$$

and such that the space $M - \bigcup U_\alpha$ has dimension strictly smaller than k. Then for any differential k-form we set

$$\int_M \omega = \sum_\alpha \int_{U_\alpha} \omega.$$

Of course, this definition seems to depend on our choice of open sets, but we can show (though we choose not to) that:

Lemma 6.6.2 *The value of $\int_M \omega$ is independent of choice of set $\{U_\alpha\}$.*

While in practice the above summation could be infinite, in which case questions of convergence must arise, in practice this is rarely a problem.

6.7 Stokes' Theorem

We now come to the goal of this chapter:

Theorem 6.7.1 (Stokes' Theorem) *Let M be an oriented k-dimensional manifold in $\mathbf{R}^n$ with boundary ∂M, a smooth (k-1)-dimensional manifold with orientation induced from the orientation of M. Let ω be a differential (k-1)-form. Then*

$$\int_M \mathrm{d}\omega = \int_{\partial M} \omega.$$

This is a sharp quantitative version of the intuition:

Average of a function on boundary = Average of derivative on interior.

This single theorem includes as special cases the classical results of the Divergence Theorem, Green's Theorem and the vector-calculus Stokes' Theorem.

We will explicitly prove Stokes' Theorem only in the special case that M is a unit cube in $\mathbf{R}^k$ and when

$$\omega = f(x_1,\ldots,x_k)\mathrm{d}x_2 \wedge \ldots \wedge \mathrm{d}x_k.$$

After proving this special case, we will sketch the main ideas behind the proof for the general case.

Proof in unit cube case: Here

$$M = \{(x_1,\ldots,x_k) : \text{ for each } i, 0 \leq x_i \leq 1\}.$$

The boundary ∂M of this cube consists of $2k$ unit cubes in $\mathbf{R}^{k-1}$. We will be concerned with the two boundary components

$$S_1 = \{(0,x_2,\ldots,x_k) \in M\}$$

and

$$S_2 = \{(1,x_2,\ldots,x_k) \in M\}.$$

For $\omega = f(x_1,\ldots,x_k)\mathrm{d}x_2 \wedge \ldots \wedge \mathrm{d}x_k$, we have

$$\mathrm{d}\omega = \sum \frac{\partial f}{\partial x_i}\mathrm{d}x_i \wedge \mathrm{d}x_2 \wedge \ldots \wedge \mathrm{d}x_k,$$

$$= \frac{\partial f}{\partial x_1}\mathrm{d}x_1 \wedge \mathrm{d}x_2 \wedge \ldots \wedge \mathrm{d}x_k,$$

since it is always the case that $\mathrm{d}x_j \wedge \mathrm{d}x_j = 0$.

Now to integrate $\mathrm{d}\omega$ along the unit cube M. We choose our orientation preserving parametrizing map to be the identity map. Then

$$\int_M \mathrm{d}\omega = \int_0^1 \cdots \int_0^1 \frac{\partial f}{\partial x_1}\mathrm{d}x_1 \cdots \mathrm{d}x_k.$$

By the Fundamental Theorem of Calculus we can do the first integral, to get

$$\int_M \mathrm{d}\omega = \int_0^1 \cdots \int_0^1 f(1,x_2,\ldots,x_k)\mathrm{d}x_2 \cdots \mathrm{d}x_k$$

$$- \int_0^1 \cdots \int_0^1 f(0,x_2,\ldots,x_k)\mathrm{d}x_2 \cdots \mathrm{d}x_k.$$

Now to look at the integral $\int_{\partial M} \omega$. Since $\omega = f(x_1, \ldots, x_k) dx_2 \wedge \ldots \wedge dx_k$, the only parts of the integral along the boundary that will not be zero will be along S_1 and S_2, both of which are unit cubes in $\mathbf{R}^{k-1}$, with coordinates given by $x_2, \ldots, x_k$. They will have opposite orientations though. (This can be seen in the example for when M is a square in the plane; then S_1 is the bottom of the square and S_2 is the top of the square. Note how the orientations on S_1 and S_2 induced from the the orientation of the square are indeed opposite.)

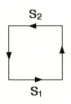

Then

$$\int_{\partial M} \omega = \int_{C_1} \omega + \int_{C_2} \omega$$

$$= \int_0^1 \cdots \int_0^1 -f(0, x_2, \ldots, x_k) dx_2 \cdots dx_k$$

$$+ \int_0^1 \cdots \int_0^1 f(1, x_2, \ldots, x_k) dx_2 \cdots dx_k,$$

which we have just shown to equal to $\int_M d\omega$, as desired. $\square$

Now to sketch a false general proof for a manifold M in $\mathbf{R}^n$. We will use that the above argument for a unit cube can be used in a similar fashion for any cube. Also, any general differential $(k-1)$-form will look like:

$$\omega = \sum f_I dx_I,$$

where each I is a (k-1)-tuple from $(1, \ldots, n)$.

Divide M into many small cubes. Adjacent cubes' boundaries will have opposite orientation.

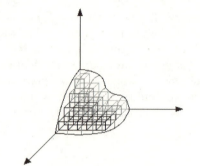

Then

$$\int_M d\omega \approx \text{Sum over the cubes} \int_{\text{little cube}} d\omega$$

$$= \text{Sum over the cubes} \int_{\partial(\text{little cube})} \omega$$

$$\approx \int_{\partial(M)} \omega.$$

The last approximation is from the fact that since the adjacent boundaries of the cubes have opposite orientations, they will cancel out. The only boundary parts that remain are those pushed out against the boundary of M itself. The final step would be to show that as we take more and more little cubes, we can replace the above approximations by equalities.

It must be noted that M cannot be split up into this union of cubes. Working around this difficultly is non-trivial.

6.8   Books

An excellent recent book is Hubbard and Hubbard's *Vector Calculus, Linear Algebra, and Differential Forms: A Unified Approach* [64], which contains a wealth of information, putting differential forms in the context of classical vector calculus and linear algebra. Spivak's *Calculus on Manifolds* [103] is for many people the best source. It is short and concise (in many ways the opposite of Spivak's leisurely presentation of ϵ and δ real analysis in [102]). Spivak emphasizes that the mathematical work should be done in getting the right definitions so that the theorems (Stokes' Theorem in particular) follow easily. Its briefness, though, makes it possibly not the best introduction. Fleming's *Functions of Several Variables* [37] is also a good introduction.

6.9 Exercises

1. Justify why it is reasonable for shuffles to indeed be called shuffles. (Think in terms of shuffling a deck of cards.)

2. In $\mathbf{R}^3$, let dx, dy and dz denote the three elementary 1-forms. Using the definition of the wedge product, show that

$$(dx \wedge dy) \wedge dz = dx \wedge (dy \wedge dz)$$

and that these are equal to the elementary 3-form $dx \wedge dy \wedge dz$.

3. Prove that for any differential k-form ω, we have

$$d(d\omega) = 0.$$

4. In $\mathbf{R}^n$, let dx and dy be one-forms. Show that

$$dx \wedge dy = -dy \wedge dx.$$

5. Prove Theorem 6.3.1.

6. Show that the map

$$\omega_{n-k}(\omega_k) = T(\omega_{n-k} \wedge \omega_k),$$

with $T : \bigwedge^n \mathbf{R^n} \to \mathbf{R}$ as defined in the chapter, provides a linear map from $\bigwedge^{n-k} \mathbf{R^n}$ to the dual space $\bigwedge^k \mathbf{R^{n*}}$.

7. Prove that the unit sphere S^2 in $\mathbf{R}^3$ is a two-dimensional manifold, using each of the three definitions.

8. Consider the rectangle

with opposite sides identified. Show first why this is a torus

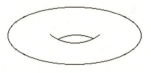

and then why it is a two-manifold.

9. The goal of this problem is to show that real projective space is a manifold. On $\mathbf{R}^{n+1} - 0$, define the equivalence relation

$$(x_0, x_1, \ldots, x_n) \sim (\lambda x_0, \lambda x_1, \ldots, \lambda x_n)$$

for any nonzero real number λ. Define real projective n-space by

$$\mathbf{P}^n = \mathbf{R}^{(n+1)} - (0)/ \sim .$$

Thus, in projective three-space, we identify $(1, 2, 3)$ with $(2, 4, 6)$ and with $(-10, -20, -30)$ but not with $(2, 3, 1)$ or $(1, 2, 5)$. In $\mathbf{P}^n$, we denote the equivalence class containing $(x_0, \ldots, x_n)$ by the notation $(x_0 : \ldots : x_n)$. Thus the point in $\mathbf{P}^3$ corresponding to $(1, 2, 3)$ is denoted by $(1 : 2 : 3)$. Then in $\mathbf{P}^3$, we have $(1 : 2 : 3) = (2 : 4 : 6) \neq (1 : 2 : 5)$. Define

$$\phi_0 : \mathbf{R}^n \to \mathbf{P}^n$$

by

$$\phi_0(u_1, \ldots, u_n) = (1 : u_1 : \ldots : u_n),$$

define

$$\phi_1 : \mathbf{R}^n \to \mathbf{P}^n$$

by

$$\phi_1(u_1, \ldots, u_n) = (u_1 : 1 : u_2 : \ldots : u_n),$$

etc., all the way up to a defining a map ϕ_n. Show that these maps can be used to make $\mathbf{P}^n$ into an n-dimensional manifold.

10. Show that the Stokes' Theorem of this chapter has as special cases:

a. the Fundamental Theorem of Calculus. (Note that we need to use the Fundamental Theorem of Calculus to prove Stokes' Theorem; thus we cannot actually claim that the Fundamental Theorem of Calculus is a mere corollary to Stokes' Theorem.)

b. Green's Theorem.

c. the Divergence Theorem.

d. the Stokes' Theorem of Chapter Five.

Chapter 7

Curvature for Curves and Surfaces

Basic Objects:	Curves and surfaces in space
Basic Goal:	Calculating curvatures

Most of high school mathematics is concerned with straight lines and planes. There is of course far more to geometry than these flat objects. Classically differential geometry is concerned with how curves and surfaces bend and twist in space. The word "curvature" is used to denote the various measures of twisting that have been discovered.

Unfortunately, the calculations and formulas to compute the different types of curvature are quite involved and messy, but whatever curvature is, it should be the case that the curvature of a straight line and of a plane must be zero, that the curvature of a circle (and of a sphere) of radius r should be the same at every point and that the curvature of a small radius circle (or sphere) should be greater than the curvature of a larger radius circle (or sphere) (which captures the idea that it is easier to balance on the surface of the earth than on a bowling ball).

The first introduction to curvature-type ideas is usually in calculus. While the first derivative gives us tangent line (and thus linear) information, it is the second derivative that measures concavity, a curvature-type measurement. Thus we should expect to see second derivatives in curvature calculations.

7.1 Plane Curves

We will describe a plane curve via a parametrization:

$$r(t) = (x(t), y(t))$$

and thus as a map

$$r : \mathbf{R} \to \mathbf{R}^2.$$

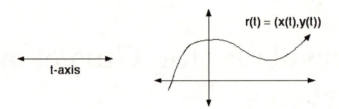

The variable t is called the *parameter* (and is frequently thought of as time). An actual plane curve can be parametrized in many different ways. For example,

$$r_1(t) = (\cos(t), \sin(t))$$

and

$$r_2(t) = (\cos(2t), \sin(2t))$$

both describe a unit circle. Any calculation of curvature should be independent of the choice of parametrization. There are a couple of reasonable ways to do this, all of which can be shown to be equivalent. We will take the approach of always fixing a canonical parametrization (the arc length parametrization). This is the parametrization $r : [a, b] \to \mathbf{R}$ such that the arc length of the curve is just $b - a$. Since the arc length is

$$\int_a^b \sqrt{\left(\frac{dx}{ds}\right)^2 + \left(\frac{dy}{ds}\right)^2} \, ds,$$

we need $\sqrt{\left(\frac{dx}{ds}\right)^2 + \left(\frac{dy}{ds}\right)^2} = 1$. Thus for the arc length parametrization, the length of the tangent vector must always be one:

$$|\mathbf{T}(s)| = \left|\frac{dr}{ds}\right| = \left|\left(\frac{dx}{ds}, \frac{dy}{ds}\right)\right| = \sqrt{\left(\frac{dx}{ds}\right)^2 + \left(\frac{dy}{ds}\right)^2} = 1.$$

Back to the question of curvature. Consider a straight line

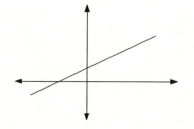

Note that each point of this line has the same tangent line.
Now consider a circle:

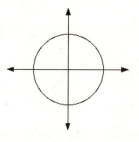

Here the tangent vectors' directions are constantly changing. This leads to the idea of trying to define curvature as a measure of the change in the direction of the tangent vectors. To measure a rate of change we need to use a derivative. This leads to:

Definition 7.1.1 *For a plane curve parametrized by arc length*

$$r(s) = (x(s), y(s)),$$

define the principal curvature κ *at a point on the curve to be the length of the derivative of the tangent vector with respect to the parameter s, i.e.,*

$$\kappa = \left| \frac{d\mathbf{T}(s)}{ds} \right|.$$

Consider the straight line $r(s) = (as + b, cs + d)$, where a, b, c and d are constants. The tangent vector is:

$$\mathbf{T}(s) = \frac{dr}{ds} = (a, c).$$

Then the curvature will be

$$\kappa = \left| \frac{d\mathbf{T}(s)}{ds} \right| = |(0,0)| = 0,$$

as desired.

Now consider a circle of radius a centered at the origin; an arc length parametrization is

$$r(s) = \left(a\cos\left(\frac{s}{a}\right), a\sin\left(\frac{s}{a}\right)\right),$$

giving us that the curvature is

$$
\begin{aligned}
\kappa &= \left|\frac{d\mathbf{T}(s)}{ds}\right| \\
&= \left|\left(-\frac{1}{a}\cos\left(\frac{s}{a}\right), -\frac{1}{a}\sin\left(\frac{s}{a}\right)\right)\right| \\
&= \sqrt{\frac{1}{a^2}\cos^2\left(\frac{s}{a}\right) + \frac{1}{a^2}\sin^2\left(\frac{s}{a}\right)} \\
&= \frac{1}{a}.
\end{aligned}
$$

Thus this definition of curvature does indeed agree with the intuitions about lines and circles that we initially desired.

7.2   Space Curves

Here the situation is more difficult; there is no single number that will capture curvature. Since we are interested in space curves, our parametrizations will have the form:

$$r(s) = (x(s), y(s), z(s)).$$

As in last section, we normalize by assuming that we have parametrized by arc length, i.e.,

$$
\begin{aligned}
|\mathbf{T}(s)| &= \left|\frac{dr}{ds}\right| = \left|\left(\frac{dx}{ds}, \frac{dy}{ds}, \frac{dz}{ds}\right)\right| \\
&= \sqrt{\left(\frac{dx}{ds}\right)^2 + \left(\frac{dy}{ds}\right)^2 + \left(\frac{dz}{ds}\right)^2} \\
&= 1.
\end{aligned}
$$

Again we start with calculating the rate of change in the direction of the tangent vector.

Definition 7.2.1 *For a space curve parametrized by arc length*

$$r(s) = (x(s), y(s), z(s)),$$

define the principal curvature κ *at a point to be the length of the derivative of the tangent vector with respect to the parameter s, i.e.,*

$$\kappa = \left| \frac{d\mathbf{T}(s)}{ds} \right|.$$

The number κ is one of the numbers that captures curvature. Another is the torsion, but before giving its definition we need to do some preliminary work.

Set

$$\mathbf{N} = \frac{1}{\kappa} \frac{d\mathbf{T}}{ds}.$$

The vector $\mathbf{N}$ is called the *principal normal vector*. Note that it has length one. More importantly, as the following proposition shows, this vector is perpendicular to the tangent vector $\mathbf{T}(s)$.

Proposition 7.2.1

$$\mathbf{N} \cdot \mathbf{T} = 0$$

at all points on the space curve.

Proof: Since we are using the arc length parametrization, the length of the tangent vector is always one, which means

$$\mathbf{T} \cdot \mathbf{T} = 1.$$

Thus

$$\frac{d}{ds}(\mathbf{T} \cdot \mathbf{T}) = \frac{d}{ds}(1) = 0.$$

By the product rule we have

$$\frac{d}{ds}(\mathbf{T} \cdot \mathbf{T}) = \mathbf{T} \cdot \frac{d\mathbf{T}}{ds} + \frac{d\mathbf{T}}{ds} \cdot \mathbf{T} = 2\mathbf{T} \cdot \frac{d\mathbf{T}}{ds}.$$

Then

$$\mathbf{T} \cdot \frac{d\mathbf{T}}{ds} = 0.$$

Thus the vectors $\mathbf{T}$ and $\frac{d\mathbf{T}}{ds}$ are perpendicular. Since the principal normal vector $\mathbf{N}$ is a scalar multiple of the vector $\frac{d\mathbf{T}}{ds}$, we have our result. $\square$

Set

$$\mathbf{B} = \mathbf{T} \times \mathbf{N},$$

a vector that is called the *binormal vector*. Since both $\mathbf{T}$ and $\mathbf{N}$ have length one, $\mathbf{B}$ must also be a unit vector. Thus at each point of the curve we have three mutually perpendicular unit vectors $\mathbf{T}$, $\mathbf{N}$ and $\mathbf{B}$. The torsion will be a number associated to the rate of change in the direction of the binormal $\mathbf{B}$, but we need a proposition before the definition can be given.

Proposition 7.2.2 *The vector $\frac{d\mathbf{B}}{ds}$ is a scalar multiple of the principal normal vector $\mathbf{N}$.*

Proof: We will show that $\frac{d\mathbf{B}}{ds}$ is perpendicular to both $\mathbf{T}$ and $\mathbf{B}$, meaning that $\frac{d\mathbf{B}}{ds}$ must point in the same direction as $\mathbf{N}$. First, since $\mathbf{B}$ has length one, by the same argument as in the previous proposition, just replacing all of the $\mathbf{T}$s by $\mathbf{B}$s, we get that $\frac{d\mathbf{B}}{ds} \cdot \mathbf{B} = 0$.

Now

$$
\begin{aligned}
\frac{d\mathbf{B}}{ds} &= \frac{d}{ds}(\mathbf{T} \times \mathbf{N}) \\
&= (\frac{d\mathbf{T}}{ds} \times \mathbf{N}) + (\mathbf{T} \times \frac{d\mathbf{N}}{ds}) \\
&= (\kappa\mathbf{N} \times \mathbf{N}) + (\mathbf{T} \times \frac{d\mathbf{N}}{ds}) \\
&= (\mathbf{T} \times \frac{d\mathbf{N}}{ds}).
\end{aligned}
$$

Thus $\frac{d\mathbf{B}}{ds}$ must be perpendicular to the vector $\mathbf{T}$. $\square$

Definition 7.2.2 *The* torsion *of a space curve is the number τ such that*

$$
\frac{d\mathbf{B}}{ds} = -\tau\mathbf{N}.
$$

We need now to have an intuitive understanding of what these two numbers mean. Basically, the torsion measures how much the space curve deviates from being a plane curve, while the principal curvature measures the curvature of the plane curve that the space curve wants to be. Consider the space curve

$$
r(s) = (3\cos\left(\frac{s}{3}\right), 3\sin\left(\frac{s}{3}\right), 5),
$$

which is a circle of radius three living in the plane $z = 5$. We will see that the torsion is zero. First, the tangent vector is

$$
\mathbf{T}(s) = \frac{dr}{ds} = (-\sin\left(\frac{s}{3}\right), \cos\left(\frac{s}{3}\right), 0).
$$

Then

$$
\frac{d\mathbf{T}}{ds} = (-\frac{1}{3}\cos\left(\frac{s}{3}\right), -\frac{1}{3}\sin\left(\frac{s}{3}\right), 0),
$$

which gives us that the principal curvature is $\frac{1}{3}$. The principal normal vector is

$$\mathbf{N} = \frac{1}{\kappa}\frac{d\mathbf{T}}{ds} = \left(-\cos\left(\frac{s}{3}\right), -\sin\left(\frac{s}{3}\right), 0\right).$$

Then the binormal is

$$\mathbf{B} = \mathbf{T} \times \mathbf{N} = (0, 0, 1),$$

and thus

$$\frac{d\mathbf{B}}{ds} = (0, 0, 0) = 0 \cdot \mathbf{N}.$$

The torsion is indeed zero, reflecting the fact that we are actually dealing with a plane curve disguised as a space curve.

Now consider the helix

$$r(t) = (\cos(t), \sin(t), t).$$

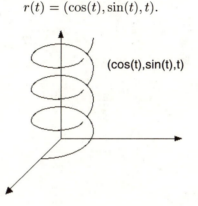

(cos(t),sin(t),t)

It should be the case that the principal curvature should be a positive constant, as the curve wants to be a circle. Similarly, the helix is constantly moving out of a plane, due to the t term in the z-coordinate. Hence the torsion should also be a nonzero constant. The tangent vector

$$\frac{dr}{dt} = (-\sin(t), \cos(t), 1)$$

does not have unique length. The arc length parametrization for this helix is simply

$$r(t) = \left(\cos\left(\frac{1}{\sqrt{2}}t\right), \sin\left(\frac{1}{\sqrt{2}}t\right), \frac{1}{\sqrt{2}}t\right).$$

Then the unit tangent vector is

$$\mathbf{T}(t) = \left(-\frac{1}{\sqrt{2}}\sin\left(\frac{1}{\sqrt{2}}t\right), \frac{1}{\sqrt{2}}\cos\left(\frac{1}{\sqrt{2}}t\right), \frac{1}{\sqrt{2}}\right).$$

The principal curvature κ is the length of the vector

$$\frac{d\mathbf{T}}{dt} = (-\frac{1}{2}\cos\left(\frac{1}{\sqrt{2}}t\right), -\frac{1}{2}\sin\left(\frac{1}{\sqrt{2}}t\right), 0).$$

Thus

$$\kappa = \frac{1}{2}.$$

Then the principal normal vector is

$$\mathbf{N}(t) = 2\frac{d\mathbf{T}}{dt} = (-\cos\left(\frac{1}{\sqrt{2}}t\right), -\sin\left(\frac{1}{\sqrt{2}}t\right), 0).$$

The binormal vector is

$$\mathbf{B} = \mathbf{T} \times \mathbf{N}$$

$$= (\frac{1}{\sqrt{2}}\sin\left(\frac{1}{\sqrt{2}}t\right), -\frac{1}{\sqrt{2}}\cos\left(\frac{1}{\sqrt{2}}t\right), \frac{1}{\sqrt{2}}).$$

The torsion τ is the length of the vector

$$\frac{d\mathbf{B}}{dt} = (\frac{1}{2}\cos\left(\frac{1}{\sqrt{2}}t\right), \frac{1}{2}\sin\left(\frac{1}{\sqrt{2}}t\right), 0),$$

and hence we have

$$\tau = \frac{1}{2}.$$

7.3   Surfaces

Measuring how tangent vectors vary worked well for understanding the curvature of space curves. A possible generalization to surfaces is to examine the variation of the tangent planes. Since the direction of a plane is determined by the direction of its normal vector, we will define curvature functions by measuring the rate of the change in the normal vector. For example, for a plane $ax + by + cz = d$, the normal at every point is the vector

$$< a, b, c > .$$

The normal vector is a constant; there is no variation in its direction. Once we have the correct definitions in place, this should provide us with the intuitively plausible idea that since the normal is not varying, the curvature must be zero.

Denote a surface by

$$X = \{(x, y, z) : f(x, y, z) = 0\}.$$

Thus we are defining our surfaces implicitly, not parametrically. The normal vector at each point of the surface is the gradient of the defining function, i.e.,

$$\mathbf{n} = \nabla f = \langle \frac{\partial f}{\partial x}, \frac{\partial f}{\partial y}, \frac{\partial f}{\partial z} \rangle.$$

Since we are interested in how the direction of the normal is changing and not in how the length of the normal is changing (since this length can be easily altered without varying the original surface at all), we normalize the defining function f by requiring that the normal $\mathbf{n}$ at every point has length one:

$$|\mathbf{n}| = 1.$$

We now have the following natural map:

Definition 7.3.1 *The* Gauss map *is the function*

$$\sigma : X \to S^2,$$

where S^2 is the unit sphere in $\mathbf{R}^2$, defined by

$$\sigma(p) = \mathbf{n}(p) = \nabla f = \langle \frac{\partial f}{\partial x}(p), \frac{\partial f}{\partial y}(p), \frac{\partial f}{\partial z}(p) \rangle.$$

As we move about on the surface X, the corresponding normal vector moves about on the sphere. To measure how this normal vector varies, we need to take the derivative of the vector-valued function σ and hence must look at the Jacobian of the Gauss map:

$$d\sigma : TX \to TS^2,$$

where TX and TS^2 denote the respective tangent planes. If we choose orthonormal bases for both of the two dimensional vector spaces TX and TS^2, we can write $d\sigma$ as a two-by-two matrix, a matrix important enough to carry its own name:

Definition 7.3.2 *The two-by-two matrix associated to the Jacobian of the Gauss map is the* Hessian.

While choosing different orthonormal bases for either TX and TS^2 will lead to a different Hessian matrix, it is the case that the eigenvalues, the trace and the determinant will remain constant (and are hence invariants of the Hessian). These invariants are what we concentrate on in studying curvature.

Definition 7.3.3 *For a surface X, the two eigenvalues of the Hessian are the* principal curvatures. *The determinant of the Hessian (equivalently the product of the principal curvatures) is the* Gaussian curvature *and the trace of the Hessian (equivalently the sum of the principal curvatures) is the* mean curvature.

We now want to see how to calculate these curvatures, in part in order to see if they agree with what our intuition demands. Luckily there is an easy algorithm that will do the trick. Start again with defining our surface X as $\{(x, y, z) : f(x, y, z) = 0\}$ such that the normal vector at each point has length one. Define the extended Hessian as

$$\tilde{H} = \begin{pmatrix} \partial^2 f/\partial x^2 & \partial^2 f/\partial x \partial y & \partial^2 f/\partial x \partial z \\ \partial^2 f/\partial x \partial y & \partial^2 f/\partial y^2 & \partial^2 f/\partial y \partial z \\ \partial^2 f/\partial x \partial z & \partial^2 f/\partial y \partial z & \partial^2 f/\partial z^2 \end{pmatrix}.$$

(Note that $\tilde{H}$ does not usually have a name.)

At a point p on X choose two orthonormal tangent vectors:

$$\mathbf{v}_1 = a_1 \frac{\partial}{\partial x} + b_1 \frac{\partial}{\partial y} + c_1 \frac{\partial}{\partial z} = \begin{pmatrix} a_1 & b_1 & c_1 \end{pmatrix}$$

$$\mathbf{v}_2 = a_2 \frac{\partial}{\partial x} + b_2 \frac{\partial}{\partial y} + c_2 \frac{\partial}{\partial z} = \begin{pmatrix} a_2 & b_2 & c_2 \end{pmatrix}.$$

Orthonormal means that we require

$$\mathbf{v}_i \cdot \mathbf{v}_j = \begin{pmatrix} a_i & b_i & c_i \end{pmatrix} \begin{pmatrix} a_j \\ b_j \\ c_j \end{pmatrix} = \delta_{ij},$$

where δ_{ij} is zero for $i \neq j$ and is one for $i = j$. Set

$$h_{ij} = \begin{pmatrix} a_i & b_i & c_i \end{pmatrix} \tilde{H} \begin{pmatrix} a_j \\ b_j \\ c_j \end{pmatrix}.$$

Then a technical argument, heavily relying on the chain rule, will yield

Proposition 7.3.1 *Coordinate systems can be chosen so that the Hessian matrix is the matrix H. Thus the principal curvatures for a surface X at a point p are the eigenvalues of the matrix*

$$H = \begin{pmatrix} h_{11} & h_{12} \\ h_{21} & h_{22} \end{pmatrix}$$

and the Gaussian curvature is $\det(H)$ and the mean curvature is $\mathrm{trace}(H)$.

We can now compute some examples. Start with a plane X given by

$$(ax + by + cz - d = 0).$$

Since all of the second derivatives of the linear function $ax + by + cz - d$ are zero, the extended Hessian is the three-by-three zero matrix, which means that the Hessian is the two-by-two zero matrix, which in turn means that the principal curvatures, the Gaussian and the mean curvature are all zero, as desired.

Now suppose $X = \{(x, y, z) : \frac{1}{2r}(x^2 + y^2 + z^2 - r^2) = 0\}$, a sphere of radius r.

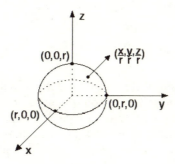

The normal is the unit vector

$$\left(\tfrac{x}{r} \quad \tfrac{y}{r} \quad \tfrac{z}{r} \right)$$

and the extended Hessian is

$$\tilde{H} = \begin{pmatrix} \frac{1}{r} & 0 & 0 \\ 0 & \frac{1}{r} & 0 \\ 0 & 0 & \frac{1}{r} \end{pmatrix} = \frac{1}{r} I.$$

Then given any two orthonormal vectors $\mathbf{v}_1$ and $\mathbf{v}_2$, we have that

$$h_{ij} = \begin{pmatrix} a_i & b_i & c_i \end{pmatrix} \tilde{H} \begin{pmatrix} a_j \\ b_j \\ c_j \end{pmatrix} = \frac{1}{r} \mathbf{v}_i \cdot \mathbf{v}_j,$$

and thus that the Hessian is the following diagonal matrix

$$H = \begin{pmatrix} \frac{1}{r} & 0 \\ 0 & \frac{1}{r} \end{pmatrix} = \frac{1}{r}I.$$

The two principal curvatures are both $\frac{1}{r}$ and are hence independent of which point is considered on the sphere, again agreeing with intuition.

For the final example, let X be a cylinder :

$$X = \{(x, y, z) : \frac{1}{2r}(x^2 + y^2 - r^2) = 0\}.$$

Since the intersection of this cylinder with any plane parallel to the xy plane is a circle of radius r, we should suspect that one of the principal curvatures should be the curvature of a circle, namely $\frac{1}{r}$. But also through each point on the cylinder there is a straight line parallel to the z-axis, suggesting that the other principal curvature should be zero. We can now check these guesses. The extended Hessian is

$$\tilde{H} = \begin{pmatrix} \frac{1}{r} & 0 & 0 \\ 0 & \frac{1}{r} & 0 \\ 0 & 0 & 0 \end{pmatrix}.$$

We can choose orthonormal tangent vectors at each point of the cylinder of the form

$$v_1 = (a \quad b \quad 0)$$

and

$$v_2 = (0 \quad 0 \quad 1).$$

Then the Hessian is the diagonal matrix

$$H = \begin{pmatrix} \frac{1}{r} & 0 \\ 0 & 0 \end{pmatrix},$$

meaning that one of the principal curvatures is indeed $\frac{1}{r}$ and the other is 0.

7.4 The Gauss-Bonnet Theorem

Curvature is not a topological invariant. A sphere and an ellipsoid are topologically equivalent (intuitively meaning that one can be continuously deformed into the other; technically meaning that there is a topological homeomorphism from one onto the other) but clearly the curvatures are different. But we can not alter curvature too much, or more accurately, if we make the appropriate curvature large near one point, it must be compensated for at other points. That is the essence of the Gauss-Bonnet Theorem, which we only state in this section.

We restrict our attention to compact orientable surfaces, which are topologically spheres, toruses, two-holed toruses, three-holed toruses, etc.

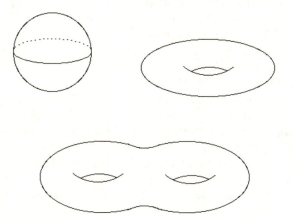

The number of holes (called the genus g) is known to be the only topological invariant, meaning that if two surfaces have the same genus, they are topologically equivalent.

Theorem 7.4.1 (Gauss-Bonnet) *For a surface X, we have*

$$\int_X Gaussian\ curvature = 2\pi(2 - 2g).$$

Thus while the Gaussian curvature is not a local topological invariant, its average value on the surface is such an invariant. Note that the left-hand side of the above equation involves analysis, while the right-hand side is topological. Equations of the form

$$\text{Analysis information} = \text{Topological information}$$

permeate modern mathematics, culminating in the Atiyah-Singer Index Formula from the mid 1960s (which has as a special case the Gauss-Bonnet Theorem). By now, it is assumed that if you have a local differential invariant, there should be a corresponding global topological invariant. The work lies in finding the correspondences.

7.5   Books

The range in texts is immense. In part this is because the differential geometry of curves and surfaces is rooted in the nineteenth century while higher dimensional differential geometry usually has quite a twentieth century feel to it. Three long time popular introductions are by do Carmo [29], Millman and Parker [85] and O'Neil [91]. A recent innovative text, emphasizing geometric intuitions is by Henderson [56]. Alfred Gray [48] has written a long book built around Mathematica, a major software package for mathematical computations. This would be a good source to see how to do actual calculations. Thorpe's text [111] is also interesting.

McLeary's *Geometry from a Differentiable Viewpoint* [84] has a lot of material in it, which is why it is also listed in the chapter on axiomatic geometry. Morgan [86] has written a short, readable account of Riemannian geometry. Then there are the classic texts. Spivak's five volumes [102] are impressive, with the first volume a solid introduction. The bible of the 1960s and 70s is *Foundations of Differential Geometry* by Kobayashi and Nomizu [74]; though fading in fashion, I would still recommend all budding differential geometers to struggle with its two volumes, but not as an introductory text.

7.6   Exercises

1. Let C be the plane curve given by $r(t) = (x(t), y(t))$. Show that the curvature at any point is

$$\kappa = \frac{x'y'' - y'x''}{((x')^2 + (y')^2)^{3/2}}.$$

(Note that the parametrization $r(t)$ is not necessarily the arc length parametrization.)

2. Let C be the plane curve given by $y = f(x)$. Show that a point $p = (x_0, y_0)$ is a point of inflection if and only if the curvature at p is zero. (Note that p is a point of inflection if $f''(x_0) = 0$.)

3. For the surface described by

$$z = x^2 + \frac{y^2}{4},$$

find the principal curvatures at each point. Sketch the surface. Does the sketch provide the same intuitions as the principal curvature calculations?

4. Consider the cone

$$z^2 = x^2 + y^2.$$

Find the image of the Gauss map. (Note that you need to make sure that the normal vector has length one.) What does this image have to say about the principal curvatures?

5. Let

$$A(t) = (a_1(t), a_2(t), a_3(t))$$

and

$$B(t) = (b_1(t), b_2(t), b_3(t))$$

be two 3-tuples of differentiable functions. Show that

$$\frac{d}{dt}(A(t) \cdot B(t)) = \frac{dA}{dt} \cdot B(t) + A(t) \cdot \frac{dB}{dt}.$$

Chapter 8

Geometry

Basic Objects:	Points and Lines in Planes
Basic Goal:	Axioms for Different Geometries

The axiomatic geometry of Euclid was the model for correct reasoning from at least as early as 300 BC to the mid 1800s. Here was a system of thought that started with basic definitions and axioms and then proceeded to prove theorem after theorem about geometry, all done without any empirical input. It was believed that Euclidean geometry correctly described the space that we live in. Pure thought seemingly told us about the physical world, which is a heady idea for mathematicians. But by the early 1800s, non-Euclidean geometries had been discovered, culminating in the early 1900s in the special and general theory of relativity, by which time it became clear that, since there are various types of geometry, the type of geometry that describes our universe is an empirical question. Pure thought can tell us the possibilities but does not appear able to pick out the correct one. (For a popular account of this development by a fine mathematician and mathematical gadfly, see Kline's *Mathematics and the Search for Knowledge* [73].)

Euclid started with basic definitions and attempted to give definitions for his terms. Today, this is viewed as a false start. An axiomatic system starts with a collection of undefined terms and a collection of relations (axioms) among these undefined terms. We can then prove theorems based on these axioms. An axiomatic system "works" if no contradictions occur. Hyperbolic and elliptic geometries were taken seriously when it was shown that any possible contradiction in them could be translated back into a contradiction in Euclidean geometry, which no one seriously believes contains a contradiction. This will be discussed in the appropriate sections of this chapter.

8.1   Euclidean Geometry

Euclid starts with twenty-three Definitions, five Postulates and five Common Notions. We will give a flavor of his language by giving a few examples of each (following Heath's translation of Euclid's *Elements* [32]; another excellent source is in Cederberg's *A Course in Modern Geometries* [17]).

For example, here is Euclid's definition of a line:

A line *is breadthless length*

and for a surface:

A surface *is that which has length and breadth only.*

While these definitions do agree with our intuitions of what these words should mean, to modern ears they sound vague.

His five Postulates would today be called axioms. They set up the basic assumptions for his geometry. For example, his fourth postulate states:

That all right angles are equal to one another.

Finally, his five Common Notions are basic assumptions about equalities. For example, his third common notion is

If equals be subtracted from equals, the remainders are equal.

All of these are straightforward, except for the infamous fifth postulate. This postulate has a different feel than the rest of Euclid's beginnings.

Fifth Postulate: *That, if a straight line falling on two straight lines makes the interior angles on the same side less than two right angles, the two straight lines, if produced indefinitely, meet on that side on which are the angles less than the two right angles.*

Certainly by looking at the picture

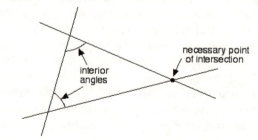

we see that this is a perfectly reasonable statement. We would be surprised if this were not true. What is troubling is that this is a basic assumption. Axioms should not be just reasonable but obvious. This is not obvious. It is also much more complicated than the other postulates, even in the superficial way that its statement requires a lot more words than the other postulates. In part, it is making an assumption about the infinite, as it states that if you extend lines further out, there will be an intersection point. A feeling of uneasiness was shared by mathematicians, starting with Euclid himself, who tried to use this postulate as little as possible.

One possible approach is to replace this postulate with another one that is more appealing, turning this troubling postulate into a theorem. There are a number of statements equivalent to the fifth postulate, but none that really do the trick. Probably the most popular is Playfair's Axiom:

Given a point off of a line, there is a unique line through the point parallel to the given line.

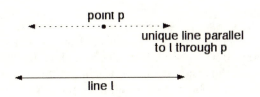

Certainly a reasonable statement. Still, it is quite bold to make this a basic assumption. It would be ideal if the fifth postulate could be shown to be a statement provable from the other axioms. The development of other geometries stemmed from the failed attempts in trying to prove the fifth postulate.

8.2 Hyperbolic Geometry

One method for showing that the fifth postulate must follow from the other axioms is to assume it is false and find a contradiction. Using Playfair's Axiom, there are two possibilities: either there are no lines through the point parallel to the given line or there are more than one line through the point parallel to the given line. These assumptions now go by the names:

Elliptic Axiom: Given a point off of a given line, there are no lines through the point parallel to the line.

This is actually just making the claim that there are no parallel lines, or that every two lines must intersect (which again seems absurd).

Hyperbolic Axiom: Given a point off of a given line, there is more than one line through the point parallel to the line.

What is meant by parallel must be clarified. Two lines are defined to be *parallel* if they do not intersect.

Geroloamo Saccheri (1667-1773) was the first to try to find a contradiction from the assumption that the fifth postulate is false. He quickly showed that if there is no such parallel line, then contradictions occurred. But when he assumed the Hyperbolic Axiom, no contradictions arose. Unfortunately for Saccheri, he thought that he had found such a contradiction and wrote a book, *Euclides ab Omni Naevo Vindicatus* (Euclid Vindicated from all Faults), that claimed to prove that Euclid was right.

Gauss (1777-1855) also thought about this problem and seems to have realized that by negating the fifth postulate, other geometries would arise. But he never mentioned this work to anybody and did not publish his results.

It was Lobatchevsky (1793-1856) and Janos Bolyai (1802-1860) who, independently, developed the first non-Euclidean geometry, now called hyperbolic geometry. Both showed, like Saccheri, that the Elliptic Axiom was not consistent with the other axioms of Euclid, and both showed, again like Saccheri, that the Hyperbolic Axiom did not appear to contradict the other axioms. Unlike Saccheri though, both confidently published their work and did not deign to find a fake contradiction.

Of course, just because you prove a lot of results and do not come up with a contradiction does not mean that a contradiction will not occur the next day. In other words, Bolyai and Lobatchevsky did not have a proof of consistency, a proof that no contradictions could ever occur. Felix Klein (1849-1925) is the main figure for finding models for different geometries that would allow for proofs of consistency, though the model we will look at was developed by Poincaré (1854-1912).

Thus the problem is how to show that a given collection of axioms forms a consistent theory, meaning that no contradiction can ever arise. The model approach will not show that hyperbolic geometry is consistent but instead show that it is as consistent as Euclidean geometry. The method is to model the straight lines of hyperbolic geometry as half circles in Euclidean geometry. Then each axiom of hyperbolic geometry will be a theorem of Euclidean geometry. The process can be reversed, so that each axiom of Euclidean geometry will become a theorem in hyperbolic geometry. Thus, if there is some hidden contradiction in hyperbolic geometry, there must also be a hidden contradiction in Euclidean geometry (a contradiction that no one believes to exist).

Now for the details of the model. Start with the upper half plane

$$H = \{(x,y) \in R^2 : y > 0\}.$$

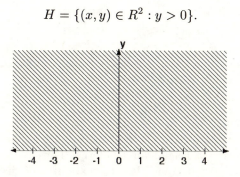

Our points will be simply the points in H. The key to our model of hyperbolic geometry is how we define straight lines. We say that a line is either a vertical line in H or a half-circle in H that intersects the x-axis perpendicularly.

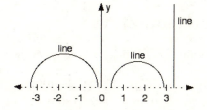

To see that this is indeed a model for hyperbolic geometry we would have to check each of the axioms. For example, we would need to check that between any two points there is a unique line (or in this case, show that for any two points in H, there is either a vertical line between them or a unique half-circle between them).

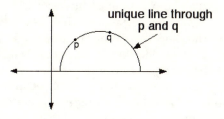

The main thing to see is that for this model the Hyperbolic Axiom is obviously true.

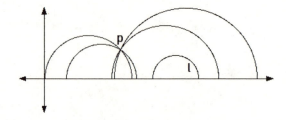

What this model allows us to do is to translate each axiom of hyperbolic geometry into a theorem in Euclidean geometry. Thus the axioms about lines in hyperbolic geometry become theorems about half-circles in Euclidean geometry. Therefore, hyperbolic geometry is as consistent as Euclidean geometry.

Further, this model shows that the fifth postulate can be assumed to be either true or false; this means that the fifth postulate is *independent* of the other axioms.

8.3 Elliptic Geometry

But what if we assume the Elliptic Axiom. Saccheri, Gauss, Bolyai and Lobatchevsky all showed that this new axiom was inconsistent with the other axioms. Could we, though, alter these other axioms to come up with another new geometry. Riemann (1826-1866) did precisely this, showing that there were two ways of altering the other axioms and thus that there were two new geometries, today called single elliptic geometry and double elliptic geometry (named by Klein). For both, Klein developed models and thus showed that both are as consistent as Euclidean geometry.

In Euclidean geometry, any two distinct points are on a unique line. Also in Euclidean geometry, a line must separate the plane, meaning that given any line l, there are at least two points off of l such that the line segment connecting the two points must intersect l.

For single elliptic geometry, we assume that *a line does not separate the plane*, in addition to the Elliptic Axiom. We keep the Euclidean assumption that any two points uniquely determine a line. For double elliptic geometry,, we need to assume that *two points can lie on more than one line*, but now keep the Euclidean assumption that a line will separate the plane. All of these sound absurd if you are thinking of straight lines as the straight lines from childhood. But under the models that Klein developed, they make sense, as we will now see.

For double elliptic geometry, our "plane" is the the unit sphere, the points are the points on the sphere and our "lines" will be the great circles on the spheres. (The *great circles* are just the circles on the sphere with greatest diameter.)

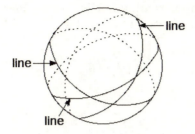

Note that any two lines will intersect (thus satisfying the Elliptic Axiom) and that while most pairs of points will uniquely define a line, points opposite to each other will lie on infinitely many lines. Thus statements about lines in double elliptic geometry will correspond to statements about great circles in Euclidean geometry.

For single elliptic geometry, the model is a touch more complicated. Our "plane" will now be the upper half-sphere, with points on the boundary circle identified with their antipodal points, i.e.,

$$\{(x,y,z) : x^2 + y^2 + z^2 = 1, z \geq 0\}/\{(x,y,0) \text{ is identified with } (-x,-y,0)\}.$$

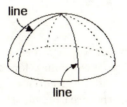

Thus the point on the boundary $(\frac{1}{\sqrt{2}}, -\frac{1}{\sqrt{2}}, 0)$ is identified with the point $(-\frac{1}{\sqrt{2}}, \frac{1}{\sqrt{2}}, 0)$. Our "lines" will be the great half-circles on the half-sphere. Note that the Elliptic Axiom is satisfied. Further, note that no line will separate the plane, since antipodal points on the boundary are identified. Thus statements in single elliptic geometry will correspond to statements about great half-circles in Euclidean geometry.

8.4 Curvature

One of the most basic results in Euclidean geometry is that the sum of the angles of a triangle is 180 degrees, or in other words, the sum of two right angles.

Recall the proof. Given a triangle with vertices P, Q and R, by Playfair's Axiom there is a unique line through R parallel to the line spanned by P and Q. By results on alternating angles, we see that the the angles α, β and γ must sum to that of two right angles.

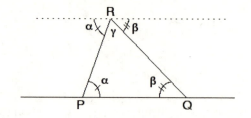

Note that we needed to use Playfair's axiom. Thus this result will not necessarily be true in non-Euclidean geometries. This seems reasonable if we look at the picture of a triangle in the hyperbolic upper half-plane and of a triangle on the sphere of double elliptic geometry.

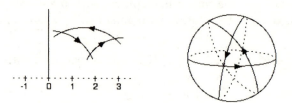

What happens is that in hyperbolic geometry the sums of the angles of a triangle are less than 180 degrees while, for elliptic geometries, the sum of the angles of a triangle will be greater than 180 degrees. It can be shown that the smaller that the area of the triangle is, the closer the sum of the triangle's angles will be to 180 degrees. This in turn is linked to the Gaussian curvature. It is the case (though it is not obvious) that methods of measuring distance (i.e., metrics) can be chosen so that the different types of geometry will have different Gaussian curvatures. More precisely, the Gaussian curvature of the Euclidean plane will be zero, of the hyperbolic plane will be -1 and of the elliptic planes will be 1. Thus differential geometry and curvature are linked to the axiomatics of different geometries.

8.5 Books

One of the best popular books in mathematics of all time is Hilbert and Cohn-Vossens' *Geometry and the Imagination* [58]. All serious students should study this book carefully. One of the 1900s best geometers (someone who actually researched in areas that nonmathematicians would recognize as geometry), Coxeter, wrote a great book, *Introduction to Geometry* [23]. More standard, straightforward texts on various types of geometry are by Gans [44], Cederberg [17] and Lang and Murrow [81] . Robin Hartshorne's *Geometry: Euclid and Beyond* [55] is an interesting recent book. Also, McLeary's *Geometry from a Differentiable Viewpoint* [84] is a place to see both non-Euclidean geometries and the beginnings of differential geometry.

8.6 Exercises

1. This problem gives another model for hyperbolic geometry. Our points will be the points in the open disc:

$$D = \{(x, y) : x^2 + y^2 < 1\}.$$

The lines will be the arcs of circles that intersect perpendicularly the boundary of D. Show that this model satisfies the Hyperbolic Axiom.

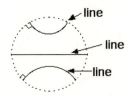

2. Show that the model in problem 1 and the upper half plane model are equivalent, if, in the upper half plane, we identify all points at infinity to a single point.

3. Give the analogue of Playfair's Axiom for planes in space.

4. Develop the idea of the upper half space so that if P is a "plane" and p is a point off of this plane, then there are infinitely many planes containing p that do not intersect the plane P.

5. Here is another model for single elliptic geometry. Start with the unit disc

$$D = \{(x, y) : x^2 + y^2 \le 1\}.$$

Identify antipodal points on the boundary. Thus identify the point (a, b) with the point $(-a, -b)$, provided that $a^2 + b^2 = 1$. Our points will be the points of the disc, subject to this identification on the boundary.

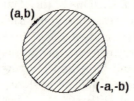

Lines will in this model be Euclidean lines, provided they start and end at
antipodal points. Show that this model describes a single elliptic geometry.
6. Here is still another model for single elliptic geometry. Let our points
be lines through the origin in space. Our lines in this geometry will be
planes through the origin in space. (Note that two lines through the origin
do indeed span a unique plane.) Show that this model describes a single
elliptic geometry.
7. By looking at how a line through the origin in space intersects the top
half of the unit sphere

$$\{(x, y, z) : x^2 + y^2 + z^2 = 1 \text{ and } z \geq 0\},$$

show that the model given in problem 6 is equivalent to the model for single
elliptic geometry given in the text.

Chapter 9

Complex Analysis

Basic Object:	The complex numbers
Basic Map:	Analytic functions
Basic Goal:	Equivalences of analytic functions

Complex analysis in one variable studies a special type of function (called analytic or holomorphic) mapping complex numbers to themselves. There are a number of seemingly unrelated but equivalent ways for defining an analytic function. Each has its advantages; all should be known.

We will first define analyticity in terms of a limit (in direct analogy with the definition of a derivative for a real-valued function). We will then see that this limit definition can also be captured by the Cauchy-Riemann equations, an amazing set of partial differential equations. Analyticity will then be described in terms of relating the function with a particular path integral (the Cauchy Integral Formula). Even further, we will see that a function is analytic if and only if it can be locally written in terms of a convergent power series. We will then see that an analytic function, viewed as a map from $\mathbf{R}^2$ to $\mathbf{R}^2$, must preserve angles (which is what the term *conformal* means), provided that the function has a nonzero derivative. Thus our goal is:

Theorem 9.0.1 *Let $f : U \to \mathbf{C}$ be a function from an open set U of the complex numbers to the complex numbers. The function $f(z)$ is said to be* analytic *if it satisfies any of the following equivalent conditions:*
a) For all $z_0 \in U$,

$$\lim_{z \to z_0} \frac{f(z) - f(z_0)}{z - z_0}$$

exists. This limit is denoted by $f'(z_0)$ and is called the complex derivative.

b) The real and imaginary parts of the function f satisfy the Cauchy-Riemann equations:

$$\frac{\partial Re(f)}{\partial x} = \frac{\partial Im(f)}{\partial y}$$

and

$$\frac{\partial Re(f)}{\partial y} = -\frac{\partial Im(f)}{\partial x}.$$

c) Let σ be a counterclockwise simple loop in U such that every interior point of σ is also in U. If z_0 is any complex number in the interior of σ, then

$$f(z_0) = \frac{1}{2\pi i} \int_\sigma \frac{f(z)}{z - z_0} dz.$$

d) For any complex number z_0, there is an open neighborhood in U of z_0 in which

$$f(z) = \sum_{n=0}^{\infty} a_n (z - z_0)^n,$$

a uniformly converging series.

Further, *if f is analytic at a point z_0 and if $f'(z_0) \neq 0$, then at z_0, the function f is conformal (i.e., angle-preserving), viewed as a map from $\mathbf{R}^2$ to $\mathbf{R}^2$.*

There is a basic distinction between real and complex analysis. Real analysis studies, in essence, differentiable functions; this is not a major restriction on functions at all. Complex analysis studies analytic functions; this is a major restriction on the type of functions studied, leading to the fact that analytic functions have many amazing and useful properties. Analytic functions appear throughout modern mathematics and physics, with applications ranging from the deepest properties of prime numbers to the subtlety of fluid flow. Know this subject well.

9.1 Analyticity as a Limit

For the rest of this chapter, let U denote an open set of the complex numbers $\mathbf{C}$.

Let $f : U \to \mathbf{C}$ be a function from our open set U of the complex numbers to the complex numbers.

Definition 9.1.1 *At a point $z_0 \in U$, the function $f(z)$ is* analytic *(or* holomorphic*) if*

$$\lim_{z \to z_0} \frac{f(z) - f(z_0)}{z - z_0}$$

exists. This limit is denoted by $f'(z_0)$ and is called the derivative.

Of course, this is equivalent to the limit

$$\lim_{h \to 0} \frac{f(z_0 + h) - f(z_0)}{h}$$

existing for $h \in \mathbf{C}$.

Note that this is exactly the definition for a function $f : \mathbf{R} \to \mathbf{R}$ to be differentiable if all $\mathbf{C}$'s are replaced by $\mathbf{R}$'s. Many basic properties of differentiable functions (such as the product rule, sum rule, quotient rule, and chain rule) will immediately apply. Hence, from this perspective, there does not appear to be anything particularly special about analytic functions. But the involved limits are not limits on the real line but limits in the real plane. This extra complexity creates profound distinctions between real differentiable functions and complex analytic ones, as we will see.

Our next task is to give an example of a nonholomorphic function. We need a little notation. The complex numbers $\mathbf{C}$ form a real two dimensional vector space. More concretely, each complex number z can be written as the sum of a real and imaginary part:

$$z = x + iy.$$

The *complex conjugate* of z is

$$\bar{z} = x - iy.$$

Note that the square of the length of the complex number z as a vector in $\mathbf{R}^2$ is

$$x^2 + y^2 = z\bar{z}.$$

Keeping in tune with this notion of length, the product $z\bar{z}$ is frequently denoted by:

$$z\bar{z} = |z|^2.$$

Fix the function

$$f(z) = \bar{z} = x - iy.$$

We will see that this function is not holomorphic. The key is that in the definition we look at the limit as $h \to 0$ but h must be allowed to be any complex number. Then we must allow h to approach 0 along any path in **C**, or in other words, along any path in $\mathbf{R}^2$. We will take the limit along two different paths and see that we get two different limits, meaning that $\bar{z}$ is not holomorphic.

For convenience, let $z_0 = 0$. Let h be real valued. Then for this h we have

$$\lim_{h \to 0} \frac{f(h) - f(0)}{h - 0} = \lim_{h \to 0} \frac{h}{h} = 1.$$

Now let h be imaginary, which we label, with an abuse of notation, by hi, with h now real. Then the limit will be:

$$\lim_{hi \to 0} \frac{f(hi) - f(0)}{hi - 0} = \lim_{h \to 0} \frac{-hi}{hi} = -1.$$

Since the two limits are not equal, the function $\bar{z}$ cannot be a holomorphic function.

9.2   Cauchy-Riemann Equations

For a function $f : U \to \mathbf{C}$, we can split the image of f into its real and imaginary parts. Then, using that

$$z = x + iy = (x, y),$$

we can write $f(z) = u(z) + iv(z)$ as

$$f(x, y) = u(x, y) + iv(x, y).$$

For example, if $f(z) = z^2$, we have

$$\begin{aligned} f(z) &= z^2 \\ &= (x + iy)^2 \\ &= x^2 - y^2 + 2xyi. \end{aligned}$$

Then the real and imaginary parts of the function f will be:

$$\begin{aligned} u(x, y) &= x^2 - y^2 \\ v(x, y) &= 2xy. \end{aligned}$$

The goal of this section is to capture the analyticity of the function f by having the real-valued functions u and v satisfy a special system of partial differential equations.

Definition 9.2.1 *Real-valued functions* $u, v : U \rightarrow \mathbf{R}$ *satisfy the* Cauchy-Riemann *equations if*

$$\frac{\partial u(x,y)}{\partial x} = \frac{\partial v(x,y)}{\partial y}$$

and

$$\frac{\partial u(x,y)}{\partial y} = -\frac{\partial v(x,y)}{\partial x}.$$

Though not at all obvious, this is the most important system of partial differential equations in all of mathematics, due to its intimate connection with analyticity, described in the following theorem.

Theorem 9.2.1 *A complex-valued function* $f(x,y) = u(x,y) + iv(x,y)$ *is analytic at a point* $z_0 = x_0 + iy_0$ *if and only if the real-valued functions* $u(x,y)$ *and* $v(x,y)$ *satisfy the Cauchy-Riemann equations at* z_0.

We will show that analyticity implies the Cauchy-Riemann equations and then that the Cauchy-Riemann equations, coupled with the condition that the partial derivatives $\frac{\partial u}{\partial x}, \frac{\partial v}{\partial x}, \frac{\partial u}{\partial y}$ and $\frac{\partial v}{\partial y}$ are continuous, imply analyticity. This extra assumption requiring the continuity of the various partials is not needed, but without it the proof is quite a bit harder.

Proof: We first assume that at a point $z_0 = x_0 + iy_0$,

$$\lim_{h \to 0} \frac{f(z_0 + h) - f(z_0)}{h}$$

exists, with the limit denoted as usual by $f'(z_0)$. The key is that the number h is a complex number. Thus when we require the above limit to exist as h approaches zero, the limit must exist along any path in the plane for h approaching zero.

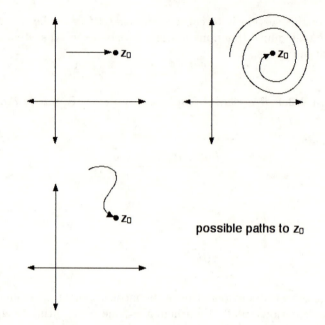

possible paths to z_0

The Cauchy-Riemann equations will follow by choosing different paths for h.

First, assume that h is real. Then

$$f(z_0 + h) = f(x_0 + h, y) = u(x_0 + h, y) + iv(x_0 + h, y).$$

By the definition of analytic function,

$$
\begin{aligned}
f'(z_0) &= \lim_{h \to 0} \frac{f(z_0 + h) - f(z_0)}{h} \\
&= \lim_{h \to 0} \frac{u(x_0 + h, y_0) + iv(x_0 + h, y_0) - (u(x_0, y_0) + iv(x_0, y_0))}{h} \\
&= \lim_{h \to 0} \frac{u(x_0 + h, y_0) - u(x_0, y_0)}{h} + i \lim_{h \to 0} \frac{v(x_0 + h, y_0) - v(x_0, y_0)}{h} \\
&= \frac{\partial u}{\partial x}(x_0, y_0) + i \frac{\partial v}{\partial x}(x_0, y_0),
\end{aligned}
$$

by the definition of partial derivatives.

Now assume that h is always purely imaginary. For ease of notation we denote h by hi, h now real. Then

$$f(z_0 + hi) = f(x_0, y_0 + h) = u(x_0, y_0 + h) + iv(x_0, y_0 + h).$$

We have, for the same complex number $f'(z_0)$ as before,

$$
\begin{aligned}
f'(z_0) &= \lim_{h \to 0} \frac{f(z_0 + ih) - f(z_0)}{ih} \\
&= \lim_{h \to 0} \frac{u(x_0, y_0 + h) + iv(x_0, y_0 + h) - (u(x_0, y_0) + iv(x_0, y_0))}{ih} \\
&= \frac{1}{i} \lim_{h \to 0} \frac{u(x_0, y_0 + h) - u(x_0, y_0)}{h} + \lim_{h \to 0} \frac{v(x_0, y_0 + h) - v(x_0, y_0)}{h} \\
&= -i \frac{\partial u}{\partial y}(x_0, y_0) + \frac{\partial v}{\partial y}(x_0, y_0),
\end{aligned}
$$

by the definition of partial differentiation and since $\frac{1}{i} = -i$.

But these two limits are both equal to the same complex number $f'(z_0)$. Hence

$$
\frac{\partial u}{\partial x} + i \frac{\partial v}{\partial x} = -i \frac{\partial u}{\partial y} + \frac{\partial v}{\partial y}.
$$

Since $\frac{\partial u}{\partial x}, \frac{\partial v}{\partial x}, \frac{\partial u}{\partial y}$, and $\frac{\partial v}{\partial y}$ are all real-valued functions, we must have

$$
\begin{aligned}
\frac{\partial u}{\partial x} &= \frac{\partial v}{\partial y} \\
\frac{\partial u}{\partial y} &= -\frac{\partial v}{\partial x},
\end{aligned}
$$

the Cauchy-Riemann equations.

Before we can prove that the Cauchy-Riemann equations (plus the extra assumption of continuity on the partial derivatives) imply that $f(z)$ is analytic, we need to describe how complex multiplication can be interpreted as a linear map from $\mathbf{R}^2$ to $\mathbf{R}^2$ (and hence as a 2×2 matrix).

Fix a complex number $a + bi$. Then for any other complex number $x + iy$, we have

$$
(a + bi)(x + iy) = (ax - by) + i(ay + bx).
$$

Representing $x + iy$ as a vector $\binom{x}{y}$ in $\mathbf{R}^2$, we see that multiplication by $a + bi$ corresponds to the matrix multiplication

$$
\begin{pmatrix} a & -b \\ b & a \end{pmatrix} \begin{pmatrix} x \\ y \end{pmatrix} = \begin{pmatrix} ax - by \\ bx + ay \end{pmatrix}.
$$

As can be seen, not all linear transformations $\binom{AB}{CD} : \mathbf{R}^2 \to \mathbf{R}^2$ correspond to multiplication by a complex number. In fact, from the above we have

Lemma 9.2.1 *The matrix*

$$
\begin{pmatrix} AB \\ CD \end{pmatrix}
$$

corresponds to multiplication by a complex number $a + bi$ if and only if $A = D = a$ and $B = -C = -b$.

Now we can return to the other direction of the theorem. First write our function $f : \mathbf{C} \to \mathbf{C}$ as a map $f : \mathbf{R}^2 \to \mathbf{R}^2$ by

$$f(x,y) = \begin{pmatrix} u(x,y) \\ v(x,y) \end{pmatrix}.$$

As described in Chapter Three, the Jacobian of f is the unique matrix

$$Df = \begin{pmatrix} \frac{\partial u}{\partial x}(x_0, y_0) & \frac{\partial u}{\partial y}(x_0, y_0) \\ \frac{\partial v}{\partial x}(x_0, y_0) & \frac{\partial v}{\partial y}(x_0, y_0) \end{pmatrix}$$

satisfying

$$\lim_{\substack{x \to x_0 \\ y \to y_0}} \frac{\left| \begin{pmatrix} u(x,y) \\ v(x,y) \end{pmatrix} - \begin{pmatrix} u(x_0,y_0) \\ v(x_0,y_0) \end{pmatrix} - Df \cdot \begin{pmatrix} x - x_0 \\ y - y_0 \end{pmatrix} \right|}{|(x - x_0, y - y_0)|} = 0.$$

But the Cauchy-Riemann equations, $\frac{\partial u}{\partial x} = \frac{\partial v}{\partial y}$ and $\frac{\partial u}{\partial y} = -\frac{\partial v}{\partial x}$, tell us that this Jacobian represents multiplication by a complex number. Call this complex number $f'(z_0)$. Then, using that $z = x + iy$ and $z_0 = x_0 + iy_0$, we can rewrite the above limit as

$$\lim_{z \to z_0} \frac{|f(z) - f(z_0) - f'(z_0)(z - z_0)|}{|z - z_0|} = 0.$$

This must also hold without the absolute value signs and hence

$$\begin{aligned}
0 &= \lim_{z \to z_0} \frac{f(z) - f(z_0) - f'(z_0)(z - z_0)}{z - z_0} \\
&= \lim_{z \to z_0} \frac{f(z) - f(z_0)}{z - z_0} - f'(z_0).
\end{aligned}$$

Thus

$$f'(z_0) = \lim_{z \to z_0} \frac{f(z) - f(z_0)}{z - z_0}$$

will always exist, meaning that the function $f : \mathbf{C} \to \mathbf{C}$ is analytic. $\square$

9.3 Integral Representations of Functions

Analytic functions can also be defined in terms of path integrals about closed loops in **C**. This means that we will be writing analytic functions as integrals, which is what is meant by the term *integral representation*. We will see that for a closed loop σ,

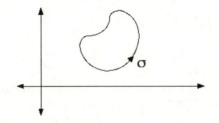

the values of an analytic function on interior points are determined from the values of the function on the boundary, which places strong restrictions on what analytic functions can be. The consequences of this integral representation of analytic functions range from the beginnings of homology theory to the calculation of difficult real-valued integrals (using residue theorems).

We first need some preliminaries on path integrals and Green's Theorem. Let σ be a path in our open set U. In other words, σ is the image of a differentiable map

$$\sigma : [0, 1] \to U.$$

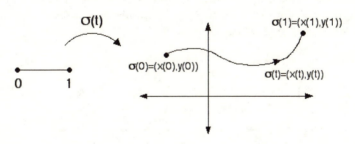

Writing $\sigma(t) = (x(t), y(t))$, with x denoting the real coordinate of **C** and y the imaginary coordinate, we have:

Definition 9.3.1 *If $P(x, y)$ and $Q(x, y)$ are real-valued functions defined on an open subset U of $\mathbf{R}^2 = \mathbf{C}$, then*

$$\int_\sigma P\mathrm{d}x + Q\mathrm{d}y = \int_0^1 P(x(t), y(t))\frac{\mathrm{d}x}{\mathrm{d}t}\mathrm{d}t + \int_0^1 Q(x(t), y(t))\frac{\mathrm{d}y}{\mathrm{d}t}\mathrm{d}t.$$

If $f : U \to \mathbf{C}$ is a function written as

$$f(z) = f(x,y) = u(x,y) + iv(x,y) = u(z) + iv(z),$$

then

Definition 9.3.2 *The* path integral $\int_\sigma f(z)\mathrm{d}z$ *is defined by*

$$
\begin{aligned}
\int_\sigma f(z)\mathrm{d}z &= \int_\sigma \big(u(x,y) + iv(x,y)\big)(\mathrm{d}x + i\mathrm{d}y) \\
&= \int_\sigma \big(u(x,y) + iv(x,y)\big)\mathrm{d}x + \int_\sigma \big(iu(x,y) - v(x,y)\big)\mathrm{d}y.
\end{aligned}
$$

The goal of this section is to see that these path integrals have a number of special properties when the function f is analytic.

A path σ is a *closed loop* in U if there is a parametrization $\sigma : [0,1] \to U$ with $\sigma(0) = \sigma(1)$.

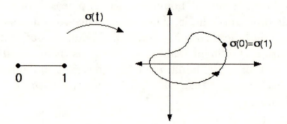

Note that we are using the same symbol for the actual path and for the parametrization function. The loop is *simple* if $\sigma(t) \neq \sigma(s)$, for all $s \neq t$, except for when t or s is zero or one.

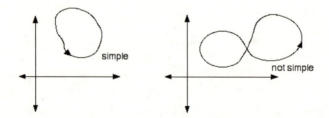

We will require all of our simple loops to be parametrized so that they are counterclockwise around their interior. For example, the unit circle is a counterclockwise simple loop, with parametrization

$$\sigma(t) = (\cos(2\pi t), \sin(2\pi t)).$$

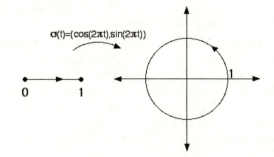

We will be interested in the path integrals of analytic functions around counterclockwise simple loops. Luckily, there are two key, easy examples that demonstrate the general results. Both of these examples will be integrals about the unit circle. Consider the function $f : \mathbf{C} \to \mathbf{C}$ defined by

$$f(z) = z = x + iy.$$

Then

$$
\begin{aligned}
\int_{\sigma} f(z)\mathrm{d}z &= \int_{\sigma} z\mathrm{d}z \\
&= \int_{\sigma} (x + iy)(\mathrm{d}x + i\mathrm{d}y) \\
&= \int_{\sigma} (x + iy)\mathrm{d}x + \int_{\sigma} (xi - y)\mathrm{d}y \\
&= \int_0^1 (\cos(2\pi t) + i\sin(2\pi t))\frac{\mathrm{d}}{\mathrm{d}t}\cos(2\pi t)\mathrm{d}t \\
&\quad + \int_0^1 (i\cos(2\pi t) - \sin(2\pi t))\frac{\mathrm{d}}{\mathrm{d}t}\sin(2\pi t)\mathrm{d}t \\
&= 0,
\end{aligned}
$$

when the integral is worked out.

On the other hand, consider the function $f(z) = \frac{1}{z}$. On the unit circle we have $|z|^2 = z\overline{z} = 1$ and hence $\frac{1}{z} = \overline{z}$. Then

$$\int_{\sigma} f(z)\mathrm{d}z = \int_{\sigma} \frac{\mathrm{d}z}{z} = \int_{\sigma} \overline{z}\mathrm{d}z = \int (\cos(2\pi t) - i\sin(2\pi t))(\mathrm{d}x + i\mathrm{d}y)$$

$$= 2\pi i,$$

when the calculation is performed. We will soon see that the reason that the path integral $\int_\sigma \frac{dz}{z}$ equals $2\pi i$ for the unit circle is that the function $\frac{1}{z}$ is not well-defined in the interior of the circle (namely at the origin). Otherwise the integral would be zero, as in the first example. Again, though not at all apparent, these are the two key examples.

The following theorems will show that the path integral of an analytic function about a closed loop will always be zero if the function is also analytic on the interior of the loop.

We will need, though, Green's Theorem:

Theorem 9.3.1 (Green's Theorem) *Let σ be a counterclockwise simple loop in $\mathbf{C}$ and Ω its interior. If $P(x,y)$ and $Q(x,y)$ are two real-valued differentiable functions, then*

$$\int_\sigma P dx + Q dy = \int \int_\Omega \left(\frac{\partial Q}{\partial x} - \frac{\partial P}{\partial y} \right) dx dy.$$

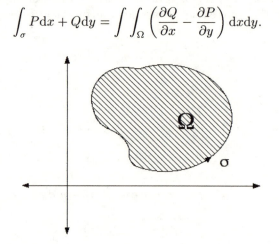

The proof is exercise 5 in Chapter Five.

Now on to Cauchy's Theorem:

Theorem 9.3.2 (Cauchy's Theorem) *Let σ be a counterclockwise simple loop in an open set U such that every point in the interior of σ is contained in U. If $f : U \to \mathbf{C}$ is an analytic function, then*

$$\int_\sigma f(z) dz = 0.$$

Viewing the path integral $\int_\sigma f(z) dz$ as some sort of average of the values of $f(z)$ along the loop σ, this theorem is stating the average value is zero for an analytic f. By the way, this theorem is spectacularly false for most functions, showing that those that are analytic are quite special.

Proof: (under the additional hypothesis, which can be removed with some work, that the complex derivative $f'(z)$ is continuous).

Write $f(z) = u(z) + iv(z)$, with $u(z)$ and $v(z)$ real-valued functions. Since $f(z)$ is analytic we know that the Cauchy-Riemann equations hold:

$$\frac{\partial u}{\partial x} = \frac{\partial v}{\partial y}$$

and

$$-\frac{\partial u}{\partial y} = \frac{\partial v}{\partial x}.$$

Now

$$
\begin{aligned}
\int_\sigma f(z)\mathrm{d}z &= \int_\sigma (u + iv)(\mathrm{d}x + i\mathrm{d}y) \\
&= \int_\sigma (u\mathrm{d}x - v\mathrm{d}y) + i\int_\sigma (u\mathrm{d}y + v\mathrm{d}x) \\
&= \int\int_\Omega \left(-\frac{\partial v}{\partial x} - \frac{\partial u}{\partial y}\right)\mathrm{d}x\mathrm{d}y + i\int\int_\Omega \left(\frac{\partial u}{\partial x} - \frac{\partial v}{\partial y}\right)\mathrm{d}x\mathrm{d}y,
\end{aligned}
$$

by Green's Theorem, where as before Ω denotes the interior of the closed loop σ. But this path integral must be zero by the Cauchy-Riemann equations. $\square$

Note that while the actual proof of Cauchy's Theorem was short, it used two major earlier results, namely the equivalence of the Cauchy-Riemann equations with analyticity and Green's Theorem.

This theorem is at the heart of all integral-type properties for analytic functions. For example, this theorem leads (nontrivially) to the following, which we will not prove:

Theorem 9.3.3 *Let $f : U \to \mathbf{C}$ be analytic in an open set U and let σ and $\hat\sigma$ be two simple loops so that σ can be continuously deformed to $\hat\sigma$ in U (i.e., σ and $\hat\sigma$ are homotopic in U). Then*

$$\int_\sigma f(z)\mathrm{d}z = \int_{\hat\sigma} f(z)\mathrm{d}z.$$

Intuitively, two loops are homotopic in a region U if one can be continuously deformed into the other within U. Thus

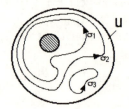

σ_1 and σ_2 are homotopic to each other in the region U but not to σ_3 in this region (though all three are homotopic to each other in $\mathbf{C}$). The technical definition is:

Definition 9.3.3 *Two paths σ_1 and σ_2 are homotopic in a region U if there is a continuous map*

$$T : [0,1] \times [0,1] \to U$$

with

$$T(t,0) = \sigma_1(t)$$

and

$$T(t,1) = \sigma_2(t).$$

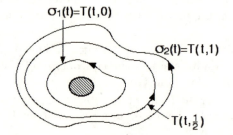

In the statement of Cauchy's Theorem, the requirement that all of the points in the interior of the closed loop σ be in the open set U can be restated as requiring that the loop σ is homotopic to a point in U.

We also need the notion of simply connected. A set U in $\mathbf{C}$ is *simply connected* if every closed loop in U is homotopic in U to a single point. Intuitively, U is simply connected if U contains the interior points of every closed loop in U. For example, the complex numbers $\mathbf{C}$ is simply connected, but $\mathbf{C}-(0,0)$ is not simply connected, since $\mathbf{C}-(0,0)$ does not contain the unit disc, even though it does contain the unit circle.

We will soon need the following slight generalization of Cauchy's Theorem:

Proposition 9.3.1 *Let U be a simply connected open set in $\mathbf{C}$. Let $f : U \to \mathbf{C}$ be analytic except possibly at a point z_0 but continuous everywhere. Let σ be any counterclockwise simple loop in U. Then*

$$\int_\sigma f(z)\mathrm{d}z = 0.$$

The proof is similar to that of Cauchy's Theorem; the extension is that we have to guarantee that all still works even if the point z_0 lies on the loop σ.

All of these lead to:

Theorem 9.3.4 (Cauchy Integral Formula) *Let $f : U \to \mathbf{C}$ be analytic on a simply connected open set U in $\mathbf{C}$ and let σ be a counterclockwise simple loop in U. Then for any point z_0 in the interior of σ, we have*

$$f(z_0) = \frac{1}{2\pi i}\int_\sigma \frac{f(z)}{z - z_0}\mathrm{d}z.$$

The meaning of this theorem is that the value of the analytic function f at any point in the interior of a region can be obtained by knowing the values of f on the boundary curve.

Proof: Define a new function $g(z)$ by setting

$$g(z) = \frac{f(z) - f(z_0)}{z - z_0},$$

when $z \neq z_0$ and setting

$$g(z) = f'(z_0)$$

when $z = z_0$.

Since $f(z)$ is analytic at z_0, by definition we have

$$f'(z_0) = \lim_{z \to z_0} \frac{f(z) - f(z_0)}{z - z_0},$$

meaning that the new function $g(z)$ is continuous everywhere and analytic everywhere except for possibly at z_0.

Then by the last theorem we have $\int_\sigma g(z)\mathrm{d}z = 0$. Thus

$$0 = \int_\sigma \frac{f(z) - f(z_0)}{z - z_0}\mathrm{d}z = \int_\sigma \frac{f(z)}{z - z_0}\mathrm{d}z - \int_\sigma \frac{f(z_0)}{z - z_0}\mathrm{d}z.$$

Then

$$\int_\sigma \frac{f(z)}{z - z_0} \mathrm{d}z \; = \; \int_\sigma \frac{f(z_0)}{z - z_0} \mathrm{d}z$$

$$= \; f(z_0) \int_\sigma \frac{1}{z - z_0} \mathrm{d}z,$$

since $f(z_0)$ is just a fixed complex number. But this path integral is just our desired $2\pi i f(z_0)$, by direct calculation, after deforming our simple loop σ to a circle centered at z_0. $\square$

In fact, the converse is also true.

Theorem 9.3.5 *Let σ be a counterclockwise simple loop and $f : \sigma \to \mathbf{C}$ any continuous function on the loop σ. Extend the function f to the interior of the loop σ by setting*

$$f(z_0) = \frac{1}{2\pi i} \int_\sigma \frac{f(z)}{z - z_0} \mathrm{d}z$$

for points z_0 in the interior. Then $f(z)$ is analytic on the interior of σ. Further, f is infinitely differentiable with

$$f^k(z_0) = \frac{k!}{2\pi i} \int_\sigma \frac{f(z)}{(z - z_0)^{k+1}} \mathrm{d}z.$$

Though a general proof is in most books on complex analysis, we will only sketch why the derivative $f'(z_0)$ is capable of being written as the path integral

$$\frac{1}{2\pi i} \int_\sigma \frac{f(z)}{(z - z_0)^2} \mathrm{d}z.$$

For ease of notation, we write

$$f(z) = \frac{1}{2\pi i} \int_\sigma \frac{f(w)}{w - z} \mathrm{d}w.$$

Then

$$f'(z) \; = \; \frac{\mathrm{d}}{\mathrm{d}z} f(z)$$

$$= \; \frac{\mathrm{d}}{\mathrm{d}z} \left(\frac{1}{2\pi i} \int_\sigma \frac{f(w)}{w - z} \mathrm{d}w \right)$$

$$= \; \frac{1}{2\pi i} \int_\sigma \frac{\mathrm{d}}{\mathrm{d}z} \left(\frac{f(w)}{w - z} \right) \mathrm{d}w$$

$$= \; \frac{1}{2\pi i} \int_\sigma \frac{f(w)}{(w - z)^2} \mathrm{d}w$$

as desired.

Note that in this theorem we are not assuming that the original function $f : \sigma \to \mathbf{C}$ was analytic. In fact the theorem is saying that any continuous function on a simple loop can be used to define an analytic function on the interior. The reason that this can only be called a sketch of a proof was that we did not justify the pulling of the derivative $\frac{d}{dz}$ inside of the integral.

9.4 Analytic Functions as Power Series

Polynomials $a_n z^n + a_{n-1} z^{n-1} + \cdots + a_0$ are great functions to work with. In particular they are easy to differentiate and to integrate. Life would be easy if all we ever had to be concerned with were polynomials. But this is not the case. Even basic functions such as e^z, $\log(z)$ and the trig functions are just not polynomials. Luckily though, all of these functions are analytic, which we will see in this section means that they are almost polynomials, or more accurately, glorified polynomials, which go by the more common name as power series. In particular the goal of this section is to prove:

Theorem 9.4.1 *Let U be an open set in $\mathbf{C}$. A function $f : U \to \mathbf{C}$ is analytic at z_0 if and only if in a neighborhood of z_0, $f(z)$ is equal to a uniformly convergent power series, i.e.,*

$$f(z) = \sum_{n=0}^{\infty} a_n (z - z_0)^n.$$

Few functions are equal to uniformly convergent power series (these "glorified polynomials"). Thus we will be indeed showing that an analytic function can be described as such a glorified polynomial.

Note that if

$$\begin{aligned}
f(z) &= \sum_{n=0}^{\infty} a_n (z - z_0)^n \\
&= a_0 + a_1 (z - z_0) + a_2 (z - z_0)^2 + \cdots,
\end{aligned}$$

we have that

$$\begin{aligned}
f(z_0) &= a_0, \\
f'(z_0) &= a_1, \\
f^{(2)}(z_0) &= 2a_2, \\
&\vdots \\
f^{(k)}(z_0) &= k!a_k.
\end{aligned}$$

Thus, if $f(z) = \sum_{n=0}^{\infty} a_n(z - z_0)^n$, we have

$$f(z) = \sum_{n=0}^{\infty} \frac{f^{(n)}(z_0)}{n!}(z - z_0)^n,$$

the function's Taylor series. In other words, the above theorem is simply stating that an analytic function is equal to its Taylor series.

We first show that any uniformly convergent power series defines an analytic function by reviewing quickly some basic facts about power series and then sketching a proof.

Recall the definition of uniform convergence, given in Chapter Three.

Definition 9.4.1 *Let U be a subset of the complex numbers $\mathbf{C}$. A sequence of functions, $f_n : A \to \mathbf{C}$, converges uniformly to a function $f : U \to \mathbf{C}$ if given any $\epsilon > 0$, there is some positive integer N such that for all $n \geq N$,*

$$|f_n(z) - f(z)| < \epsilon,$$

for all points z in U.

In other words, we are guaranteed that eventually all the functions $f_n(z)$ will fall within any ϵ-tube about the limit function $f(z)$.

The importance of uniform convergence for us is the following theorem, which we will not prove here:

Theorem 9.4.2 *Let the sequence $\{f_n(z)\}$ of analytic functions converge uniformly on an open set U to a function $f : U \to \mathbf{C}$. Then the function $f(z)$ is also analytic and the sequence of derivatives $(f_n'(z))$ will converge pointwise to the derivative $f'(z)$ on the set U.*

Now that we have a definition for a sequence of functions to converge uniformly, we can make sense out of what it would mean for a series of functions to converge uniformly, via translating series statements into sequence statements using the partial sums of the series.

Definition 9.4.2 *A series $\sum_{n=0}^{\infty} a_n(z - z_0)^n$, for complex numbers a_n and z_0, converges uniformly in an open set U of the complex numbers $\mathbf{C}$ if the sequence of polynomials $\{\sum_{n=0}^{N} a_n(z - z_0)^n\}$ converges uniformly in U.*

By the above theorem and since polynomials are analytic, we can conclude that if

$$f(z) = \sum_{n=0}^{\infty} a_n(z - z_0)^n$$

is a uniformly convergent series, then the function $f(z)$ is analytic.

Now to sketch why any analytic function can be written as a uniformly convergent power series. The Cauchy Integral Formula from last section will be critical.

Start with a function f which is analytic about a point z_0. Choose a simple loop σ about z_0. By the Cauchy Integral Formula,

$$f(z) = \frac{1}{2\pi i} \int_\sigma \frac{f(w)}{w - z}\,dw,$$

for any z inside σ.

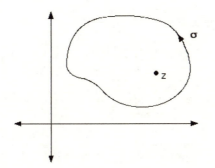

Knowing that the geometric series is

$$\sum_{n=0}^{\infty} r^n = \frac{1}{1 - r},$$

for $|\,r\,| < 1$, we see that, for all w and z with $|z - z_0| < |w - z_0|$, we have

$$\begin{aligned}
\frac{1}{w - z} &= \frac{1}{w - z_0} \cdot \frac{1}{1 - \frac{z - z_0}{w - z_0}} \\
&= \frac{1}{w - z_0} \cdot \sum_{n=0}^{\infty} \left(\frac{z - z_0}{w - z_0} \right)^n.
\end{aligned}$$

Restrict the numbers w to lie on the loop σ. Then for those complex numbers z with $|z - z_0| < |w - z_0|$,

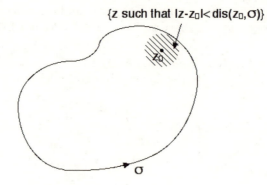

{z such that |z-z₀|< dis(z₀,σ)}

we have

$$
\begin{aligned}
f(z) &= \frac{1}{2\pi i} \int_\sigma \frac{f(w)}{w-z}\,dw \\
&= \frac{1}{2\pi i} \int_\sigma \frac{f(w)}{w-z_0} \cdot \frac{1}{1-\frac{z-z_0}{w-z_0}}\,dw \\
&= \frac{1}{2\pi i} \int_\sigma \frac{f(w)}{w-z_0} \sum_{n=0}^{\infty} \left(\frac{z-z_0}{w-z_0}\right)^n dw \\
&= \frac{1}{2\pi i} \int_\sigma \sum_{n=0}^{\infty} \frac{f(w)}{w-z_0} \left(\frac{z-z_0}{w-z_0}\right)^n dw \\
&= \frac{1}{2\pi i} \sum_{n=0}^{\infty} \int_\sigma \frac{f(w)}{w-z_0} \left(\frac{z-z_0}{w-z_0}\right)^n dw \\
&= \frac{1}{2\pi i} \sum_{n=0}^{\infty} (z-z_0)^n \int_\sigma \frac{f(w)}{(w-z_0)^{n+1}}\,dw \\
&= \sum_{n=0}^{\infty} \frac{f^{(n)}(z_0)}{n!}(z-z_0)^n,
\end{aligned}
$$

a convergent power series.

Of course the above is not quite rigorous, since we did not justify the switching of the integral with the sum. It follows, nontrivially, from the fact that the series $\sum_{n=0}^{\infty} \left(\frac{z-z_0}{w-z_0}\right)^n$ converges uniformly.

Note that we have also used the Cauchy Integral Formula, namely that

$$
f^{(n)}(z_0) = \frac{n!}{2\pi i} \int_\sigma \frac{f(w)}{(w-z_0)^{n+1}}\,dw.
$$

9.5 Conformal Maps

We now want to show that analytic functions are also quite special when one looks at the geometry of maps from $\mathbf{R}^2$ to $\mathbf{R}^2$. After defining conformal maps (the technical name for those maps that preserve angles), we will show that an analytic function will be conformal at those points where its derivative is nonzero. This will be seen to follow almost immediately from the Cauchy-Riemann equations.

Before defining angle-preserving, we need a description for the angle between curves. Let

$$\sigma_1 : [-1, 1] \to \mathbf{R}^2,$$

with $\sigma_1(t) = (x_1(t), y_1(t))$, and

$$\sigma_2 : [-1, 1] \to \mathbf{R}^2,$$

with $\sigma_2(t) = (x_2(t), y_2(t))$, be two differentiable curves in the plane which intersect at

$$\sigma_1(0) = \sigma_2(0).$$

The angle between the two curves is defined to be the angle between the curves' tangent vectors.

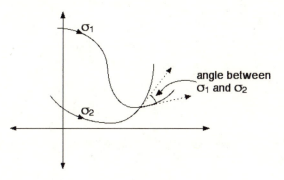

Thus we are interested in the dot product between the tangent vectors of the curves:

$$
\begin{aligned}
\frac{d\sigma_1}{dt} \cdot \frac{d\sigma_2}{dt} &= \left(\frac{dx_1}{dt}, \frac{dy_1}{dt} \right) \cdot \left(\frac{dx_2}{dt}, \frac{dy_2}{dt} \right) \\
&= \frac{dx_1}{dt} \frac{dx_2}{dt} + \frac{dy_1}{dt} \frac{dy_2}{dt}.
\end{aligned}
$$

Definition 9.5.1 *A function $f(x, y) = (u(x, y), v(x, y))$ will be* conformal *at a point (x_0, y_0) if the angle between any two curves intersecting at (x_0, y_0) is preserved, i.e., the angle between curves σ_1 and σ_2 is equal to the angle between the image curves $f(\sigma_1)$ and $f(\sigma_2)$.*

Thus

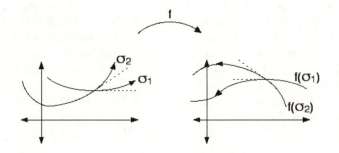

is conformal while

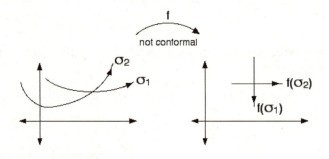

is not.

Theorem 9.5.1 *An analytic function $f(z)$ whose derivative at the point z_0 is not zero will be conformal at z_0.*

Proof : The tangent vectors are transformed under the map f by multiplying them by the two-by-two Jacobian matrix for f. Thus we want to show that multiplication by the Jacobian preserves angles. Writing f in its real and imaginary parts, with $z = x + iy$, as

$$f(z) = f(x, y) = u(x, y) + iv(x, y),$$

the Jacobian of f at the point $z_0 = (x_0, y_0)$ will be

$$Df(x_0, y_0) = \begin{pmatrix} \frac{\partial u}{\partial x}(x_0, y_0) & \frac{\partial u}{\partial y}(x_0, y_0) \\ \frac{\partial v}{\partial x}(x_0, y_0) & \frac{\partial v}{\partial y}(x_0, y_0) \end{pmatrix}.$$

But the function f is analytic at the point z_0 and hence the Cauchy-Riemann equations

$$\frac{\partial u}{\partial x}(x_0, y_0) = \frac{\partial v}{\partial y}(x_0, y_0)$$

$$-\frac{\partial u}{\partial y}(x_0, y_0) = \frac{\partial v}{\partial x}(x_0, y_0)$$

hold, allowing us to write the Jacobian as

$$Df(x_0, y_0) = \left(\begin{array}{cc} \frac{\partial u}{\partial x}(x_0, y_0) & \frac{\partial u}{\partial y}(x_0, y_0) \\ -\frac{\partial u}{\partial y}(x_0, y_0) & \frac{\partial u}{\partial x}(x_0, y_0) \end{array} \right).$$

Note that the columns of this matrix are orthogonal (i.e., their dot product is zero). This alone shows that the multiplication by the Jacobian will preserve angle. We can also show this by explicitly multiplying the Jacobian by the two tangent vectors $\frac{d\sigma_1}{dt}$ and $\frac{d\sigma_2}{dt}$ and then checking that the dot product between $\frac{d\sigma_1}{dt}$ and $\frac{d\sigma_2}{dt}$ is equal to the dot product of the image tangent vectors. $\square$

This proof uses the Cauchy-Riemann equation approach to analyticity. A more geometric (and unfortunately a more vague) approach is to look carefully at the requirement for

$$\lim_{h \to 0} \frac{f(z_0 + h) - f(z_0)}{h}$$

to exist, no matter what path is chosen for h to approach zero. This condition must place strong restrictions on how the function f alters angles.

This also suggests how to approach the converse. It can be shown (though we will not) that a conformal function f must satisfy either the limit for analyticity

$$\lim_{h \to 0} \frac{f(z_0 + h) - f(z_0)}{h}$$

or that the limit holds for the conjugate function $\bar{f}$

$$\lim_{h \to 0} \frac{\bar{f}(z_0 + h) - \bar{f}(z_0)}{h},$$

where the conjugate function of $f(z) = u(z) + iv(z)$ is

$$\bar{f}(z) = u(z) - iv(z).$$

9.6   The Riemann Mapping Theorem

Two domains D_1 and D_2 are said to be *conformally equivalent* if there is a one-to-one onto conformal map

$$f : D_1 \to D_2.$$

If such a function f exists, then its inverse function will also be conformal. Since conformal basically means that f is analytic, if two domains are conformally equivalent, then it is not possible to distinguish between them using the tools from complex analysis. Considering that analytic functions are special among functions, it is quite surprising that there are clean results for determining when two domains are conformally equivalent. The main result is:

Theorem 9.6.1 (Riemann Mapping Theorem) *Two simply connected domains, neither of which are equal to* **C**, *are conformally equivalent.*

   (Recall that a domain is *simply connected* if any closed loop in the domain is homotopic to a point in the domain, or intuitively, if every closed loop in the domain can be continuously shrunk to a point.) Frequently this result is stated as: for any simply connected domain D that is not equal to **C**, there is a conformal one-to-one onto map from D to the unit disc. Thus the domain

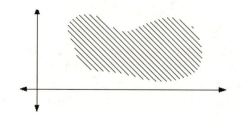

is conformally equivalent to

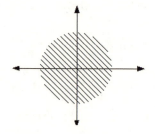

The Riemann Mapping Theorem, though, does not produce for us the desired function f. In practice, it is an art to find the conformal map. The standard approach is to first find conformal maps from each of the domains to the unit disc. Then, to conformally relate the two domains, we just compose various maps to the disc and inverses of maps to the disc.

For example, consider the right half plane

$$D - \{z \in \mathbf{C} : \operatorname{Re}(z) > 0\}.$$

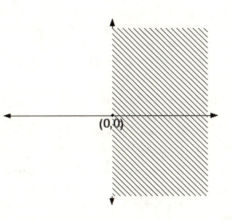

The function

$$f(z) = \frac{1 - z}{1 + z}$$

provides our conformal map from D to the unit disc. This can be checked by showing that the boundary of D, the y-axis, maps to the boundary of the unit disc. In this case, the inverse to f is f itself.

The Riemann Mapping Theorem is one reason why complex analysts spend so much time studying the function theory of the disc, as knowledge about the disc can be easily translated to knowledge about any simply connected domain.

In several complex variables theory, all is much more difficult, in large part because there is no higher dimensional analogue of the Riemann Mapping Theorem. There are many simply connected domains in $\mathbf{C}^n$ that are not conformally equivalent.

9.7   Several Complex Variables: Hartog's Theorem

Let $f(z_1, \ldots, z_n)$ be a complex-valued function of n complex variables. We say that f is *holomorphic* (or *analytic*) in several variables if $f(z_1, \ldots, z_n)$ is holomorphic in each variable z_i separately. Although many of the basic results for one variable analytic functions can be easily carried over to the several variable case, the subjects are profoundly different. These differences start with Hartog's Theorem, which is the subject of this section.

Consider the one-variable function $f(z) = \frac{1}{z}$. This function is holomorphic at all points except at the origin, where it is not even defined. It is thus easy to find a one-variable function that is holomorphic except for at one point. But what about the corresponding question for holomorphic functions of several variables? Is there a function $f(z_1, \ldots, z_n)$ that is holomorphic everywhere except at an isolated point? Hartog's theorem is that no such function can exist.

Theorem 9.7.1 (Hartog's Theorem) *Let U be an open connected region in $\mathbf{C}^n$ and let V be a compact connected set contained in U. Then any function $f(z_1, \ldots, z_n)$ that is holomorphic on $U - V$ can be extended to a holomorphic function that is defined on all of U.*

This certainly includes the case when V is an isolated point. Before sketching a proof for a special case of this theorem, consider the following question that is now quite natural, namely, is there a natural condition on open connected sets U so that there will exist holomorphic functions on U that cannot be extended to a larger open set. Such sets U are called *domains of holomorphy*. Hartog's Theorem says that regions like $U -$ (isolated point) are not domains of holomorphy. In fact, a clean criterion does exist and involves geometric conditions on the boundary of the open set U (technically, the boundary must be pseudoconvex). Hartog's Theorem opens up a whole new world of phenomena for several complex variables.

One way of thinking about Hartog's Theorem is in considering the function $\frac{f(z_1, \ldots, z_n)}{g(z_1, \ldots, z_n)}$, where both f and g are holomorphic, as a possible counterexample. If we can find a holomorphic function g that has a zero at an isolated point or even on a compact set, then Hartog's Theorem will be false. Since Hartog's Theorem is indeed a theorem, an analytic function in more than one variable cannot have a zero at an isolated point. In fact, the study of the zero locus $g(z_1, \ldots, z_n) = 0$ leads to much of algebraic and analytic geometry.

Now to sketch a proof of Hartog's Theorem, subject to simplifying assumptions that U is the polydisc

$$U = \{(z, w) : |z| < 1, |w| < 1\}$$

and that V is the isolated point $(0,0)$. We will also use the fact that if two functions that are holomorphic on an open connected region U are equal on an open subset of U, then they are equal on all of U. (The proof of this fact is similar to the corresponding result in one-variable complex analysis, which can be shown to follow from exercise three at the end of this chapter.)

Let $f(z,w)$ be a function that is holomorphic on $U - (0,0)$. We want to extend f to be a holomorphic function on all of U. Consider the sets $z = c$, where c is a constant with $|c| < 1$. Then the set

$$(z = c) \bigcap (U - (0,0))$$

is an open disc of radius one if $c \neq 0$ and an open disc punctured at the origin if $c = 0$. Define a new function by setting

$$F(z,w) = \frac{1}{2\pi i} \int_{|v| = \frac{1}{2}} \frac{f(z,v)}{v - w} dv.$$

This will be our desired extension. First, the function F is defined at all points of U, including the origin. Since the z variable is not varying in the integral, we have by Cauchy's Integral Formula that $F(z,w)$ is holomorphic in the w variable. Since the original function f is holomorphic with respect to the z variable, we have that F is holomorphic with respect to z; thus F is holomorphic on all of U. But again by Cauchy's Integral Formula, we have that $F = f$ when $z \neq 0$. Since the two holomorphic functions are equal on an open set of U, then we have equality on $U - (0,0)$.

The general proof of Hartog's Theorem is similar, namely to reduce the problem to slicing the region U into a bunch of discs and punctured discs and then using Cauchy's Integral Formula to create the new extension.

9.8 Books

Since complex analysis has many applications, there are many beginning textbooks, each emphasizing different aspects of the subject. An excellent introduction is in Marsden and Hoffman's *Basic Complex Analysis* [83]. Palka's *An Introduction to Complex Function Theory* [92] is also an excellent text. (I first learned complex analysis from Palka.) A recent beginning book is Greene and Krantz' *Function Theory of One Complex Variable* [49]. For a rapid fire introduction, Spiegels' *Complex Variables* [101] is outstanding, containing a wealth of concrete problems.

There are a number of graduate texts in complex analysis, which do start at the beginning but then build quickly. Ahlfors' book [1] has long been the standard. It reflects the mathematical era in which it was written (the 1960s) and thus approaches the subject from a decidedly abstract point of view. Conway's *Functions of One Complex Variable* [21] has long been the prime competitor to Ahlfors for the beginning graduate student market and is also quite good. The recent book by Berenstein and Gay [8] provides a modern framework for complex analysis. A good introduction to complex analysis in several variables is Krantz' *Function Theory in Several Variables* [77].

Complex analysis is probably the most beautiful subject in undergraduate mathematics. Neither Krantz' *Complex Analysis: The Geometric Viewpoint* [78] nor Davis' *The Schwarz Function and its Applications* [25] are textbooks but both show some of the fascinating implications contained in complex analysis and are good places to see how how analytic functions can be naturally linked to other parts of mathematics.

9.9   Exercises

1. Letting $z = x + iy$, show that the function

$$f(z) = f(x, y) = y^2$$

is not analytic. Show that it does not satisfy the Cauchy Integral Formula

$$f(z_0) = \frac{1}{2\pi i} \int_\sigma \frac{f(z)}{z - z_0} dz,$$

for the case when $z_0 = 0$ and when the closed loop σ is the circle of radius one centered at the origin.

2. Find a function $f(z)$ that is not analytic, besides the function given in problem one. If you think of $f(z)$ as a function of the two variables

$$f(x, y) = u(x, y) + iv(x, y),$$

almost any choice of functions u and v will work.

3. Let $f(z)$ and $g(z)$ be two analytic functions that are equal at all points on a closed loop σ. Show that for all points z in the interior of the closed loop we have the two functions equal. As a hint, start with the assumption that $g(z)$ is the zero function and thus that $f(z)$ is zero along the loop σ. Then show that $f(z)$ must also be the zero function inside the loop.

4. Find a one-to-one onto conformal map from the unit disc $\{(x, y) : x^2 + y^2 < 1\}$ to the first quadrant of the plane $\{(x, y) : x > 0 \text{ and } y > 0\}$.

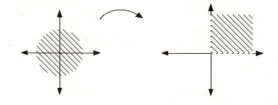

5. Let z_1, z_2 and z_3 be three distinct complex numbers. Show that we can find numbers a, b, c and d with $ad - bc = 1$ such that the map

$$T(z) = \frac{az + b}{cz + d}$$

maps z_1 to 0, z_2 to 1 and z_3 to 2. Show that the numbers a, b, c and d are uniquely determined, up to multiplication by -1.

6. Find $\int_{-\infty}^{\infty} \frac{dx}{1+x^2}$ as follows:

   a. Find

$$\int_\gamma \frac{dz}{1 + z^2},$$

where $\gamma = \gamma_1 + \gamma_2$ is the closed loop in the complex plane

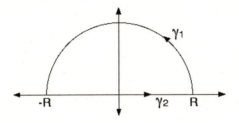

consisting of the path

$$\gamma_1 = \{Re^{\pi\theta} : 0 \le \theta\pi\}$$

and

$$\gamma_2 = \{(x,0) \in \mathbf{R}^2 : -R \le x \le R\}.$$

   b. Show that

$$\lim_{R \to \infty} \int_{\gamma_1} \frac{dz}{1 + z^2} = 0.$$

   c. Conclude with the value for $\int_{-\infty}^{\infty} \frac{dx}{1+x^2}$.

(This is a standard problem showing how to calculate hard real integrals easily. This is a hard problem if you have never used residues before; it should be straightforward if you have.)

7. The goal of this problem is to construct a conformal map from the unit sphere (minus the north pole) to the complex numbers. Consider the sphere $S^2 = \{(x,y,z) : x^2 + y^2 + z^2 = 1\}$.

   a. Show that the map

$$\pi : S^2 - (0,0,1) \to \mathbf{C}$$

defined by

$$\pi(x,y,z) = \frac{x}{1 - z} + i\frac{y}{1 - z}$$

is one-to-one, onto and conformal.

b. We can consider the complex numbers $\mathbf{C}$ as sitting inside $\mathbf{R}^3$ by mapping $x + iy$ to the point $(x, y, 0)$. Show that the above map π can be interpreted as the map that sends a point (x, y, z) on $S^2 - (0, 0, 1)$ to the point on the plane $(z = 0)$ that is the intersection of the plane with the line through (x, y, z) and $(0, 0, 1)$.

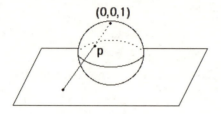

c. Justify why people regularly identify the unit sphere with $\mathbf{C} \cup \infty$.

Chapter 10

Countability and the Axiom of Choice

Basic goal: Comparing infinite sets

Both countability and the axiom of choice grapple with the elusive notions behind "infinity". While both the integers **Z** and the real numbers **R** are infinite sets, we will see that the infinity of the reals is strictly larger than the infinity of the integers. We will then turn to the Axiom of Choice, which, while straightforward and not an axiom at all for finite sets, is deep and independent from the other axioms of mathematics when applied to infinite collections of sets. Further, the Axiom of Choice implies a number of surprising and seemingly paradoxical results. For example, we will show that the Axiom of Choice forces the existence of sets of real numbers that cannot be measured.

10.1 Countability

The key is that there are different orders or magnitudes of infinity. The first step is to find the right definition for when two sets are of the same size.

Definition 10.1.1 *A set A is* finite of cardinality n *if there is a one-to-one onto function from the set $\{1, 2, 3, \cdots, n\}$ to A. The set A is* countably infinite *if there is a one-to-one onto function from the natural numbers* $\mathbf{N} = \{1, 2, 3, \ldots, \}$ *to A. A set that is either finite or countably infinite is said to be* countable. *A set A is* uncountably infinite *if it is not empty and not countable.*

For example, the set $\{a, b, c\}$ is finite with 3 elements. The more troubling and challenging examples appear in the infinite cases.

For example, the positive even numbers

$$2\mathbf{N} = \{2, 4, 6, 8, \cdots\},$$

while properly contained in the natural numbers $\mathbf{N}$, are of the same size as $\mathbf{N}$ and hence are countably infinite. An explicit one-to-one onto map

$$f : \mathbf{N} \rightarrow 2\mathbf{N}$$

is $f(n) = 2 \cdot n$. Usually this one-to-one correspondence is shown via:

The set of whole numbers $\{0, 1, 2, 3, \ldots\}$ is also countably infinite, as seen by the one-to-one onto map

$$f : \mathbf{N} \rightarrow \{0, 1, 2, 3, \ldots\}$$

given by

$$f(n) = n - 1.$$

Here the picture is

The integers $\mathbf{Z}$ are also countably infinite. The picture is while an explicit one-to-one onto function

$$f : \mathbf{N} \to \mathbf{Z}$$

is, for even n,

$$f(n) = \frac{n}{2}$$

and, for odd n,

$$f(n) = -\frac{n-1}{2}.$$

It is typical for the picture to be more convincing than the actual function.

The rationals

$$\mathbf{Q} = \{\frac{p}{q} : p, q \in \mathbf{Z}, q \neq 0\}$$

are also countably infinite. The picture for showing that the positive rationals are countably infinite is as follows:

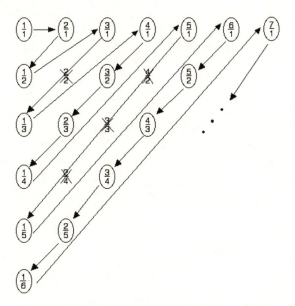

Every positive rational appears in the above array and will eventually be hit by a natural number.

In fact

Theorem 10.1.1 *Let A and B be two countably infinite sets. Then the Cartesian product $A \times B$ is also countably infinite.*

Proof: Since both A and B are in one-to-one correspondence with the natural numbers $\mathbf{N}$, all we need show is that the product $\mathbf{N} \times \mathbf{N}$ is countably infinite. For $\mathbf{N} \times \mathbf{N} = \{(n, m) : n, m \in \mathbf{N}\}$, the correct diagram is:

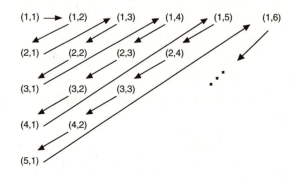

More algebraically, but less clearly, an explicit one-to-one onto map

$$f : \mathbf{N} \times \mathbf{N} \to \mathbf{N}$$

is

$$f(m, n) = \frac{(n + m - 2)(n + m - 1)}{2} + m. \quad \square$$

Note that the fact that $\mathbf{N} \times \mathbf{N}$ is the same size as $\mathbf{N}$ is of course in marked contrast to the finite case. To make this painfully obvious, consider $A = \{a, b, c\}$, a set with three elements. Then $A \times A$ is the nine element set $\{(a, a), (a, b), (a, c), (b, a), (b, b), (b, c), (c, a), (c, b), (c, c)\}$.

There are infinite sets which, in some sense, are of size strictly larger than the natural numbers. Far from being esoteric, the basic example is the set of real numbers; the reals, while certainly not finite, are also not countably infinite.

We will give the famed Cantor diagonalization argument showing that the real numbers $[0, 1] = \{x \in \mathbf{R} : 0 \le x \le 1\}$ cannot be countable.

Theorem 10.1.2 *The interval $[0, 1]$ is not countable.*

Proof : The proof is by contradiction. We assume that there is a one-to-one onto map $f : \mathbf{N} \to [0, 1]$ and then find a real number in $[0, 1]$ that is not in the image, contradicting the assumption that f is onto. We will use that every real number in $[0, 1]$ can be expressed as a decimal expansion

$$0.x_1 x_2 x_3 x_4 \ldots,$$

where each x_k is $0, 1, 2, 3, \ldots$ or 9. To make this expansion unique, we will always round up, except for the case $0.99999\ldots$ which we leave as is. Thus $0.32999\ldots$ will always be written as 0.3300.

Now let us take our assumed one-to-one correspondence $f : \mathbf{N} \to [0,1]$ and start writing down its terms. Let

$$
\begin{aligned}
f(1) &= .a_1 a_2 a_3 \cdots, \\
f(2) &= .b_1 b_2 b_3 \cdots, \\
f(3) &= .c_1 c_2 c_3 \cdots, \\
f(4) &= .d_1 d_2 d_3 \cdots, \\
f(5) &= .e_1 e_2 e_3 \cdots,
\end{aligned}
$$

and so forth. Note that the a_i, b_j, etc. are now fixed numbers between 0 and 9, given to us by the assumed one-to-one correspondence. They are not variables.

We will construct a new real number $.N_1 N_2 N_3 N_4 \ldots$ which will never appear in the above list, forcing a contradiction to the assumption that f is onto. Set

$$
N_k = \begin{cases} 4, & \text{if the } k^{th} \text{entry of } f(k) \neq 4 \\ 5, & \text{if the } k^{th} \text{entry of } f(k) = 4 \end{cases}
$$

(The choice of the numbers 4 and 5 are not important; any two integers between 0 and 9 would do just as well.)

Note that N_1 is 4 if $a_1 \neq 4$ and is 5 if $a_1 = 4$. Thus, no matter what,

$$
.N_1 N_2 N_3 \ldots \neq .a_1 a_2 \ldots = f(1).
$$

Likewise N_2 is 4 if $b_2 \neq 4$ and is 5 if $b_2 = 4$ and hence

$$
.N_1 N_2 N_3 \ldots \neq .b_2 b_2 b_3 \ldots = f(2).
$$

This continues. Since our decimal expansions are unique, and since each N_k is defined so that it is not equal to the kth term in $f(k)$, we must have that $.N_1 N_2 N_3 \cdots$ is not equal to any $f(k)$, meaning that f cannot be onto. Thus there can never be an onto function from the natural numbers to the interval $[0,1]$. Since the reals are certainly not finite, they must be uncountably infinite.

10.2 Naive Set Theory and Paradoxes

The question of what is a mathematical object was a deep source of debate in the last part of the eighteenth and first part of the nineteenth century. There has only been at best a partial resolution, caused in part by Gödel's work in logic and in part by exhaustion. Does a mathematical object exist only if an algorithm can be written that will explicitly construct the object or does it exist if the assumption of its existence leads to no contradictions, even if we can never find an example? The tension between constructive proofs versus existence proofs has in the last thirty years been eased with the development of complexity theory. The constructive camp was led by Kronecker (1823-1891), Brouwer (1881-1966) and Bishop (1928-1983). The existential camp, led by Hilbert (1862-1943), won the war, leading to most mathematicians' belief that all of mathematics can be built out of a correct set-theoretic foundation, usually believed to be an axiomatic system called Zermelo-Fraenkel plus the Axiom of Choice (for a list of those axioms, see Paul Cohen's *Set Theory and the Continuum Hypothesis* [20] Chapter II, Sections 1 and 2). This is in spite of the fact that few working mathematicians can actually write down these axioms, which certainly suggests that our confidence in our work does not stem from the axioms. More accurately, the axioms were chosen and developed to yield the results we already know to be true. In this section we informally discuss set theory and then give the famed Zermelo-Russell paradox, which shows that true care must be exercised in understanding sets.

The naive idea of a set is pretty good. Here a set is some collection of objects sharing some property. For example

$$\{n : n \text{ is an even number}\}$$

is a perfectly reasonable set. Basic operations are union, intersection and complement. We will see now how to build integers out of sets.

First for one subtlety. Given a set A, we can always form a new set, denoted by $\{A\}$, which consists of just one element, namely the set A. If A is the set of all even integers and thus containing an infinite number of elements, the set $\{A\}$ has only one element. Given a set A, we define the successor set A^+ as the union of the set A with the set $\{A\}$. Thus $x \in A^+$ if either $x \in A$ or $x = \{A\}$.

We start with the empty set $\emptyset$, the set that contains no elements. This set will correspond to the integer 0. Then we label the successor to the empty set by 1:

$$1 = \emptyset^+ = \{\emptyset\},$$

the successor to the successor of the empty set by 2:

$$2 = (\emptyset^+)^+ = \{\emptyset, \{\emptyset\}\},$$

and in general the successor to the set n by $n + 1$.

By thinking of the successor as adding by one, we can recover by recursion addition and thus in turn multiplication, subtraction and division.

Unfortunately, just naively proceeding along in this fashion will lead to paradoxes. We will construct here what appears to be a set but which cannot exist. First, note that sometimes a set can be a member of itself and sometimes not (at least if we are working in naive set theory; much of the mechanics of Zermelo-Fraenkel set theory is to prevent such nonchalant assumptions about sets). For example, the set of even numbers is not itself an even number and hence is not an element of itself. On the other hand, the set of all elements that are themselves sets with more than two elements is a member of itself. We can now define our paradoxical set. Set

$$X = \{A : A \text{ is a set that does not contain itself}\}$$
$$= \{A : A \notin A\}.$$

Is the set X an element of itself? If $X \in X$, then by the definition of X, we must have $X \notin X$, which is absurd. But if $X \notin X$, then $X \in X$, which is also silly. There are problems with allowing X to be a set. This is the Zermelo-Russell paradox

Do not think this is just a trivial little problem. Russell (1872-1970) reports in his autobiography that when he first thought of this problem he was confident it could easily be resolved, probably that night after dinner. He spent the next year struggling with it and had to change his whole method of attack on the foundations of mathematics. (Russell, with Whitehead 1861-1947), did not use set theory but instead developed type theory; type theory is abstractly no better or worse than set theory, but mathematicians base their work on the language of set theory, probably by the historical accident of World War II, which led US mathematicians to be taught by German refugees, who knew set theory, as Zermelo (1871-1953 was German.)

Do not worry too much about the definitions of set theory. You should be nervous, though, if your sets refer to themselves, as this is precisely what led to the above difficulty.

10.3   The Axiom of Choice

The axioms in set theory were chosen and developed to yield the results we already know to be true. Still, we want these axioms to be immediately obvious. Overall, this is the case. Few of the actual axioms are controversial, save for the Axiom of Choice, which states:

Axiom 10.3.1 (Axiom of Choice) *Let $\{X_\alpha\}$ be a family of nonempty sets. Then there is a set X which contains, from each set X_α, exactly one element.*

For a finite collection of sets, this is obvious and not at all axiomatic (meaning that it can be proven from other axioms). For example, let $X_1 = \{a, b\}$ and $X_2 = \{c, d\}$. Then there is certainly a set X containing one element from X_1 and one element from X_2; for example, just let $X = \{a, c\}$.

The difficulties start to arise when applying the axiom to an infinite (possibly uncountably infinite) number of sets. The Axiom of Choice gives no method for finding the set X; it just mandates the existence of X. This leads to the observation that if the Axiom of Choice is needed to prove the existence of some object, then you will never be able to actually construct that object. In other words, there will be no method to actually construct the object; it will merely be known to exist.

Another difficulty lies not in the truth of the axiom of choice but in the need to assume it as an axiom. Axioms should be clear and obvious. No one would have any difficulty with its statement if it could be proven to follow from the other axioms.

In 1939, Kurt Gödel showed that the Axiom of Choice is consistent with the other axioms. This means that using the Axiom of Choice will lead to no contradictions that were not, in some sense, already present in the other axioms. But in the early 1960s, Paul Cohen [20] showed that the Axiom of Choice was independent of the other axioms, meaning that it cannot be derived from the other axioms and hence was truly an axiom. In particular, one can assume that the Axiom of Choice is false and still be confident that no contradictions will arise.

A third difficulty with the Axiom of Choice is that it is equivalent to any number of other statements, some of which are quite bizarre. To see some of the many equivalences to the Axiom of Choice, see Howard and Rubin's *Consequences of the Axiom of Choice* [62]. One of these equivalences is the subject of the next section.

10.4 Non-measurable Sets

Warning: This section will assume a working knowledge of Lebesgue measure on the real numbers. In particular, we will need that

- If a set A is measurable, its measure $m(A)$ is equal to its outer measure $m^*(A)$.

- If $A_1, A_2, \ldots$ are disjoint sets that are measurable, then the union is measurable, with

$$m(\bigcup_{i=1}^{\infty} A_i) = \sum_{i=1}^{\infty} m(A_i).$$

This last condition corresponds to the idea that if we have two sets with lengths a and b, say, then the length of the two sets placed next to each other should be $a + b$. Also, this example closely follows the example of a nonmeasurable set in Royden's *Real Analysis* [95].

We will find a sequence of disjoint sets $A_1, A_2, \ldots$, all of which have the same outer measure and hence, if measurable, the same measure, whose union is the unit interval $[0, 1]$. Since the Lebesgue measure of the unit interval is just its length, we will have

$$1 = \sum_{i=1}^{\infty} m(A_i).$$

If each A_i is measurable, since the measures are equal, this would mean that we can add a number to itself infinitely many times and have it sum to one. This is absurd. If a series converges, then the individual terms in the series must converge to zero. Certainly they cannot all be equal.

The point of this section is that to find these sets A_i, we will need to use the Axiom of Choice. This means that we are being fairly loose with the term "find", as these sets will in no sense actually be constructed. Instead, the Axiom of Choice will allow us to claim their existence, without actually finding them.

We say that x and y in $\in [0, 1]$ are equivalent, denoted by $x \equiv y$, if $x - y$ is a rational number. It can be checked that this is an equivalence relation (see Appendix A for the basic properties of equivalence relations) and thus splits the unit interval into disjoint equivalency classes.

We now apply the Axiom of Choice to these disjoint sets. Let A be the set containing exactly one element from each of these equivalency classes. Thus the difference between any two elements of A cannot be a rational number. Note again, we do not have an explicit description of A. We have no way of knowing if a given real number is in A, but, by the Axiom of Choice, the set A does exist. In a moment we will see that A cannot be measurable.

We will now find a countable collection of disjoint sets, each with the same outer measure as the outer measure of the set A, whose union will be the unit interval. Now, since the rational numbers in $[0, 1]$ are countable, we can list all rational numbers between zero and one as $r_0, r_1, r_2, \ldots$. For convenience, assume that $r_0 = 0$. For each rational number r_i, set

$$A_i = A + r_i \,(\mathrm{mod}\, 1).$$

Thus the elements of A_i are of the form

$$a + r_i - \text{greatest integer part of } (a + r_i).$$

In particular, $A = A_0$. It is also the case that for all i

$$m^*(A) = m^*(A_i),$$

which is not hard to show, but is mildly subtle since we are not just shifting the set A by the number r_i but are then modding out by one.

We now want to show that the A_i are disjoint and cover the unit interval. First, assume that there is a number x in the intersection of A_i and A_j. Then there are numbers a_i and a_j in the set A such that

$$x = a_i + r_i \ (\mathrm{mod}\ 1) = a_j + r_j \ (\mathrm{mod}\ 1).$$

Then $a_i - a_j$ is a rational number, meaning that $a_i \equiv a_j$, which forces $i = j$. Thus if $i \neq j$, then

$$A_i \cap A_j = \emptyset.$$

Now let x be any element in the unit interval. It must be equivalent to some element a in A. Thus there is a rational number r_i in the unit interval with either

$$x = a + r_i \text{ or } a = x + r_i.$$

In either case we have $x \in A_i$. Thus the A_i are indeed a countable collection of disjoint sets that cover the unit interval. But then we have the length of the unit interval as an infinite series of the same number:

$$1 = \sum_{i=1}^{\infty} m(A_i) = \sum_{i=1}^{\infty} m(A),$$

which is impossible. Thus the set A cannot be measurable.

10.5   Gödel and Independence Proofs

In the debates about the nature of mathematical objects, all agreed that correct mathematics must be consistent (i.e., it should not be possible to both prove a statement and its converse). Eventually it was realized that most people were also implicitly assuming that mathematics was complete (meaning that any mathematical statement must ultimately be capable of being either proven or disproven). David Hilbert wanted to translate both of these goals into precise mathematical statements, each capable of rigorous proof. This attempt became known as Formalism. Unfortunately for Hilbert's school, K. Gödel (1906-1977) in 1931 destroyed any of these hopes. Gödel showed:

Any axiomatic system strong enough to include basic arithmetic must have statements in it that can be neither proven nor disproven, within the system.

Further, the example Gödel gave of a statement that could be neither proven nor disproven was that the given axiomatic system was itself consistent.

Thus in one fell swoop, Gödel showed that both consistency and completeness were beyond our grasp. Of course, no one seriously thinks that modern mathematics has within it a hidden contradiction. There are statements, though, that people care about that are not capable of being proven or disproven within Zermelo-Fraenkel set theory. The Axiom of Choice is an example of this. Such statements are said to be *independent* of the other axioms of mathematics. On the other hand, most open questions in mathematics are unlikely to be independent of Zermelo-Fraenkel set theory plus the Axiom of Choice. One exception is the question of P=NP (discussed in Chapter Sixteen), which many are now believing to be independent of the rest of mathematics.

10.6 Books

For many years the best source for getting an introduction to set theory has been Halmos' *Naive Set Theory* [53], which he wrote, in large part, to teach himself the subject. A more recent text is Moschovakis' *Notes on Set Theory* [87]. An introduction, not to set theory, but to logic is *Incompleteness Phenomenon* by Goldstern and Judah [46]. A slightly more advanced text, by a tremendous expositor, is Smullyan's *Gödel's Incompleteness Theorems* [100]. A concise, high level text is Cohen's *Set Theory and the Continuum Hypothesis* [20].

A long time popular introduction to Gödel's work has been Nagel and Newman's *Gödel's Proof* [89]. This is one of the inspirations for the amazing book of Hofstadter, *Gödel, Escher and Bach* [61]. Though not precisely a math book, it is full of ideas and should be read by everyone. Another impressive recent work is Hintikka's *Principles of Mathematics, Revisited* [60]. Here a new scheme for logic is presented. It also contains a summary of Hintikka's game-theoretic interpretation of Gödel's work.

10.7 Exercises

1. Show that the set

$$\{ax^2 + bx + c : a, b, c \in \mathbf{Q}\}$$

of all one variable polynomials of degree two with rational coefficients is countable.

2. Show that the set of all one variable polynomials with rational coefficients is countable.

3. Show that the set

$$\{a_0 + a_1 x + a_2 x^2 + \ldots : a_0, a_1, a_2, \ldots \in \mathbf{Q}\}$$

of all formal power series in one variable with rational coefficients is not countable.

4. Show that the set of all infinite sequences consisting of zeros and twos is uncountable. (This set will be used to show that the Cantor set, which will be defined in Chapter Twelve, is uncountable.)

5. In section two, the whole numbers were defined as sets. Addition by one was defined. Give a definition for addition by two and then a definition in general for whole numbers. Using this definition, show that $2 + 3 = 3 + 2$.

6.(Hard) A set S is partially ordered if there is an operation $<$ such that given any two elements x and y, we have $x < y$, $y < x$, $x = y$ or x and y have no relationship. The partial ordering is a total ordering if it must be the case that given any two elements x and y, it must be the case that $x < y$, $y < x$ or $x = y$. For example, if S is the real numbers, the standard interpretation of $<$ as less than places a total ordering on the reals. On the other hand, if S is the set of all subsets of some other set, then a partial ordering would exist if we let $<$ denote set containment. This is not a total ordering since given any two subsets, it is certainly not the case that one must be contained in the other. A partially ordered set is called a *poset*.

Let S be a poset. A chain in S is a subset of S on which the partial ordering becomes a total ordering. *Zorn's Lemma* states that if S is a poset such that every chain has an upper bound, then S contains a maximal element. Note that the upper bound to a chain need not be in the chain and that the maximal element need not be unique.

    a. Show that the Axiom of Choice implies Zorn's Lemma.

    b. Show that Zorn's Lemma implies the Axiom of Choice (this is quite a bit harder).

7. (Hard) The Hausdorff Maximal Principle states that every poset has a maximal chain, meaning a chain that is not strictly contained in any other chain. Show that the Hausdorff Maximal Principle is equivalent to the Axiom of Choice.

8. (Hard) Show that the Axiom of Choice (via the Hausdorff Maximal Principle) implies that every field is contained in an algebraically closed field. (For the definitions, see Chapter Eleven.)

Chapter 11

Algebra

Basic Objects:	Groups and rings
Basic Maps:	Group and ring homomorphisms

While current abstract algebra does indeed deserve the adjective *abstract*, it has both concrete historical roots and modern day applications. Central to undergraduate abstract algebra is the notion of a group, which is the algebraic interpretation of the geometric idea of symmetry. We can see something of the richness of groups in that there are three distinct areas that gave birth to the correct notion of an abstract group: attempts to find (more accurately, attempts to prove the inability to find) roots of polynomials, the study by chemists of the symmetries of crystals, and the application of symmetry principles to solve differential equations.

The inability to generalize the quadratic equation to polynomials of degree greater than or equal to five is at the heart of Galois Theory and involves the understanding of the symmetries of the roots of a polynomial. Symmetries of crystals involve properties of rotations in space. The use of group theory to understand the symmetries underlying a differential equation leads to Lie Theory. In all of these the idea and the applications of a group are critical.

11.1 Groups

This section presents the basic definitions and ideas of group theory.

Definition 11.1.1 *A nonempty set G that has a binary operation*

$$G \times G \to G,$$

denoted for all elements a and b in G by a · b, is a group if:

i) There is an element $e \in G$ such that $e \cdot a = a \cdot e = a$, for all a in G. (The element e is of course called the identity.)

ii) For any $a \in G$, there is an element denoted by a^{-1} such that $aa^{-1} = a^{-1}a = e$. (Naturally enough, a^{-1} is called the inverse of a.)

iii) For all $a, b, c \in G$, we have $(a \cdot b) \cdot c = a \cdot (b \cdot c)$ (i.e., we must have associativity).

Note that commutativity is not required.

Now for some examples. Let **GL(n,R)** denote the set of all $n \times n$ invertible matrices with real coefficients. Under matrix multiplication, we claim that **GL(n,R)** is a group. The identity element of course is simply the identity matrix

$$\begin{pmatrix} 1 & \cdots & 0 \\ & \ddots & \\ 0 & \cdots & 1 \end{pmatrix}.$$

The inverse of an element will be its matrix inverse. The check that matrix multiplication is associative is a long calculation. The final thing to check is to see that if A and B are invertible $n \times n$ matrices, then their product, $A \cdot B$, must be invertible. From the key theorem of linear algebra, a matrix is invertible if and only if its determinant is nonzero. Using that $\det(A \cdot B) = \det(A) \det(B)$, we have

$$\det(A \cdot B) = \det(A) \cdot \det(B) \neq 0.$$

Thus **GL(n,R)** is a group.

Note that for almost any choice of two matrices

$$A \cdot B \neq B \cdot A.$$

The group is not commutative. Geometrically, we can interpret the elements of **GL(n,R)** as linear maps on $\mathbf{R}^n$. In particular, consider rotations in three-space. These do not commute (showing this is an exercise at the end of this chapter). Rotations can be represented as invertible 3×3 matrices and hence as elements in **GL(3,R)**. If we want groups to be an algebraic method for capturing symmetry, then we will want rotations in space to form a group. Hence we cannot require groups to be commutative. (Note that rotations are associative, which is why we do require groups to be associative.)

The key examples of finite groups are the permutation groups. The permutation group, $\mathbf{S}_n$, is the set of all permutations on n distinct elements. The binary operation is composition while the identity element is the trivial permutation that permutes nothing.

To practice with the usual notation, let us look at the group of permutations on three elements:

$$\mathbf{S}_3 = \{e, (12), (13), (23), (123), (132)\}.$$

Of course we need to explain the notation. Fix an ordered triple (a_1, a_2, a_3) of numbers. Here order matters. Thus (cow, horse, dog) is different from the triple (dog, horse, cow). Each element of $\mathbf{S}_3$ will permute the ordering of the ordered triple. Specifically, the element (12) permutes (a_1, a_2, a_3) to (a_2, a_1, a_3):

$$(a_1, a_2, a_3) \overset{(12)}{\mapsto} (a_2, a_1, a_3).$$

For example, the element (12) will permute (cow, horse, dog) to the triple (horse, cow, dog). The other elements of the group $\mathbf{S}_3$ act as follows: (13) permutes (a_1, a_2, a_3) to (a_3, a_2, a_1) :

$$(a_1, a_2, a_3) \overset{(13)}{\mapsto} (a_3, a_2, a_1),$$

(23) permutes (a_1, a_2, a_3) to (a_1, a_3, a_2):

$$(a_1, a_2, a_3) \overset{(23)}{\mapsto} (a_1, a_3, a_2),$$

(123) permutes (a_1, a_2, a_3) to (a_3, a_1, a_2):

$$(a_1, a_2, a_3) \overset{(123)}{\mapsto} (a_3, a_1, a_2),$$

(132) permutes (a_1, a_2, a_3) to (a_2, a_3, a_1):

$$(a_1, a_2, a_3) \overset{(132)}{\mapsto} (a_2, a_3, a_1),$$

and of course the identity element e leave the triple (a_1, a_2, a_3) alone:

$$(a_1, a_2, a_3) \overset{(e)}{\mapsto} (a_1, a_2, a_3).$$

By composition we can multiply the permutations together, to get the following multiplication table for $\mathbf{S}_3$:

$\cdot$	e	(12)	(13)	(23)	(123)	(132)
e	e	(12)	(13)	(23)	(123)	(132)
(12)	(12)	e	(123)	(132)	(13)	(23)
(13)	(13)	(132)	e	(123)	(23)	(12)
(23)	(23)	(123)	(132)	e	(12)	(13)
(123)	(123)	(23)	(12)	(13)	(132)	e
(132)	(132)	(13)	(23)	(12)	e	(123)

Note that $\mathbf{S}_3$ is not commutative. In fact, $\mathbf{S}_3$ is the smallest possible non-commutative group. In honor of one of the founders of group theory, Niels Abel, we have:

Definition 11.1.2 *A group that is commutative is* abelian.

The integers $\mathbf{Z}$ under addition form an abelian group. Most groups are not abelian.

We want to understand all groups. Of course, this is not actually doable. Hopefully we can at least build up groups from possibly simpler, more basic groups. To start this process, we make the following definition:

Definition 11.1.3 *A nonempty subset H of G is a* subgroup *if H is itself a group, using the binary operation of G.*

For example, let

$$H = \left\{ \begin{pmatrix} a_{11} & a_{12} & 0 \\ a_{21} & a_{22} & 0 \\ 0 & 0 & 1 \end{pmatrix} : \begin{pmatrix} a_{11} & a_{12} \\ a_{21} & a_{22} \end{pmatrix} \in \mathbf{GL(2,R)}) \right\}.$$

Then H is a subgroup of the group $\mathbf{GL(3,R)}$ of invertible 3×3 matrices.

Definition 11.1.4 *Let G and $\hat{G}$ be two groups. Then a function*

$$\sigma : G \to \hat{G}$$

is a group homomorphism *if for all $g_1, g_2 \in G$,*

$$\sigma(g_1 \cdot g_2) = \sigma(g_1) \cdot \sigma(g_2).$$

For example, let $A \in \mathbf{GL(n,R)}$. Define $\sigma : \mathbf{GL(n,R)} \to \mathbf{GL(n,R)}$ by

$$\sigma(B) = A^{-1}BA.$$

Then for any two matrices $B, C \in \mathbf{GL(n,R)})$, we have

$$\sigma(BC) = A^{-1}BCA$$

$$= A^{-1}BAA^{-1}CA$$

$$= \sigma(B) \cdot \sigma(C).$$

There is a close relationship between group homomorphisms and a special class of subgroup. Before we can exhibit this, we need:

Definition 11.1.5 *Let H be a subgroup of G. The (left)* cosets *of G are all sets of the form*

$$gH = \{gh : h \in H\},$$

for $g \in G$.

This defines an equivalence class on G, with

$$g \sim \hat{g}$$

if the set gH is equal to the set $\hat{g}H$, i.e., if there is an $h \in H$ with $gh = \hat{g}$. In a natural way, the *right cosets* are the sets

$$Hg = \{hg : h \in H\},$$

which also define an equivalence relation on the group G.

Definition 11.1.6 *A subgroup H is* normal *if for all g in G, $gHg^{-1} = H$.*

Theorem 11.1.1 *Let H be a subgroup of G. The set of cosets gH, under the binary operation*

$$gH \cdot \hat{g}H = g\hat{g}H,$$

will form a group if and only if H is a normal subgroup. (This group is denoted by G/H and pronounced G mod H.)

Sketch of Proof: Most of the steps are routine. The main technical difficulty lies in showing that the binary operation

$$(gH) \cdot (\hat{g}H) = (g\hat{g}H)$$

is well defined. Hence we must show that the set $gH \cdot \hat{g}H$, which consists of the products of all elements of the set gH with all elements of the set $\hat{g}H$, is equal to the set $g\hat{g}H$. Since H is normal, we have

$$\hat{g}H(\hat{g})^{-1} = H.$$

Then as sets

$$\hat{g}H = H\hat{g}.$$

Thus

$$gH\hat{g}H = g\hat{g}H \cdot H = g\hat{g}H,$$

since $H \cdot H = H$, as H is a subgroup. The map is well defined.

The identity element of G/H is $e \cdot H$. The inverse to gH is $g^{-1}H$. Associativity follows from the associativity of the group G. $\square$

Note that in writing $gH \cdot \hat{g}H = g\hat{g}H$, one must keep in mind that H is representing every element in H and thus that H is itself not a single element.

As an application of this new group G/H, we now define the cyclic groups $\mathbf{Z}/n\mathbf{Z}$. Here our initial group is the integers $\mathbf{Z}$ and our subgroup consists of all the multiples of some fixed integer n:

$$n\mathbf{Z} = \{nk : k \in \mathbf{Z}\}.$$

Since the integers form an abelian group, every subgroup, including $n\mathbf{Z}$, is normal and thus $\mathbf{Z}/n\mathbf{Z}$ will form a group, It is common to represent each coset in $\mathbf{Z}/n\mathbf{Z}$ by an integer between 0 and $n-1$:

$$\mathbf{Z}/n\mathbf{Z} = \{0, 1, 2, \ldots, n-1\}.$$

For example, if we let $n = 6$, we have $\mathbf{Z}/6\mathbf{Z} = \{0, 1, 2, 3, 4, 5\}$. The addition table is then

+	0	1	2	3	4	5
0	0	1	2	3	4	5
1	1	2	3	4	5	0
2	2	3	4	5	0	1
3	3	4	5	0	1	2
4	4	5	0	1	2	3
5	5	0	1	2	3	4

An enjoyable exercise is proving the following critical theorem relating normal subgroups and group homomorphisms.

Theorem 11.1.2 *Let $\sigma : G \to \hat{G}$ be a group homomorphism. If*

$$ker(\sigma) = \{g \in G : \sigma(g) = \hat{e}, \text{ the identity of } \hat{G}\},$$

then $ker(\sigma)$ is a normal subgroup of G. (This subgroup $ker(\sigma)$ is called the kernel *of the map σ.)*

The study of groups is to a large extent the study of normal subgroups. By the above, this is equivalent to the study of group homomorphisms and is an example of the mid-twentieth century tack of studying an object by studying its homomorphisms.

The key theorem in finite group theory, Sylow's Theorem, links the existence of subgroups from the knowledge of the number of elements in a group.

Definition 11.1.7 *The* order *of a group G, denoted by $|G|$, is equal to the number of elements in G.*

For example, $|\mathbf{S}_3| = 6$.

Theorem 11.1.3 (Sylow's Theorem) *Let G be a finite group.*
a) Let p be a prime number. Suppose that p^α divides $|G|$. Then G has a subgroup of order p^α.

b) If p^n divides $|G|$ but p^{n+1} does not, then for any two subgroups H and $\hat{H}$ of order p^n, there is an element $g \in G$ with $gHg^{-1} = \hat{H}$.

c) If p^n divides $|G|$ but p^{n+1} does not, then the number of subgroups of order p^n is $1 + kp$, for some k a positive integer.

Proofs can be found in Herstein's *Topics in Algebra* [57], Section 2.12.

The importance lies in that we gather quite a bit of information about a finite group from merely knowing how many elements it has.

11.2 Representation Theory

Certainly one of the basic examples of groups is that of invertible $n \times n$ matrices. Representation theory studies how any given abstract group can be realized as a group of matrices. Since $n \times n$ matrices, via matrix multiplication on column vectors, are linear transformations from a vector space to itself, we can rephrase representation theory as the study of how a group can be realized as a group of linear transformations.

If V is a vector space, let $\mathbf{GL(V)}$ denote the group of linear transformations from V to itself.

Definition 11.2.1 *A* representation *of a group G on a vector space V is a group homomorphism*

$$\rho : G \to \mathbf{GL(V)}.$$

We say that ρ is a representation of G.

For example, consider the group $\mathbf{S}_3$ of permutations on three elements. There is quite a natural representation of $\mathbf{S}_3$ on three space $\mathbf{R}^3$. Let

$$\begin{pmatrix} a_1 \\ a_2 \\ a_3 \end{pmatrix} \in \mathbf{R}^3.$$

If $\sigma \in \mathbf{S}_3$, then define the map ρ by:

$$\rho(\sigma) \begin{pmatrix} a_1 \\ a_2 \\ a_3 \end{pmatrix} = \begin{pmatrix} a_{\sigma(1)} \\ a_{\sigma(2)} \\ a_{\sigma(3)} \end{pmatrix}.$$

For example, if $\sigma = (12)$, then

$$\rho(12) \begin{pmatrix} a_1 \\ a_2 \\ a_3 \end{pmatrix} = \begin{pmatrix} a_2 \\ a_1 \\ a_3 \end{pmatrix}.$$

As a matrix, we have:

$$\rho(12) = \begin{pmatrix} 0 & 1 & 0 \\ 1 & 0 & 0 \\ 0 & 0 & 1 \end{pmatrix}.$$

If $\sigma = (123)$, then since (123) permutes (a_1, a_2, a_3) to (a_3, a_1, a_2), we have

$$\rho(123) \begin{pmatrix} a_1 \\ a_2 \\ a_3 \end{pmatrix} = \begin{pmatrix} a_3 \\ a_1 \\ a_2 \end{pmatrix}.$$

As a matrix,

$$\rho(123) = \begin{pmatrix} 0 & 0 & 1 \\ 1 & 0 & 0 \\ 0 & 1 & 0 \end{pmatrix}.$$

The explicit matrices representing the other elements of $\mathbf{S}_3$ are left as an exercise at the end of the chapter.

The goal of representation theory is to find all possible representations for a given group. In order to even be able to start to make sense out of this question, we first see how to build new representations out of old.

Definition 11.2.2 *Let G be a group. Suppose we have representations of G:*

$$\rho_1 : G \to \mathbf{GL}(\mathbf{V}_1)$$

and

$$\rho_2 : G \to \mathbf{GL}(\mathbf{V}_2),$$

where V_1 and V_2 are possibly different vector spaces. Then the direct sum *representation of G on $V_1 \oplus V_2$, denoted by*

$$(\rho_1 \oplus \rho_2) : G \to \mathbf{GL}(\mathbf{V}_1) \oplus \mathbf{GL}(\mathbf{V}_2),$$

is defined for all $g \in G$ by:

$$(\rho_1 \oplus \rho_2)(g) = \rho_1(g) \oplus \rho_2(g).$$

Note that when we write out $\rho_1(g) \oplus \rho_2(g)$ as a matrix, it will be in block diagonal form.

If we want to classify representations, we should concentrate on finding those representations that are not direct sums of other representations. This leads to:

Definition 11.2.3 *A representation ρ of a group G on a nonzero vector space V is* irreducible *if there is no proper subspace W of V such that for all $g \in G$ and all $w \in W$,*

$$\rho(g)w \in W.$$

In particular if a representation is the direct sum of two other representations, it will certainly not be irreducible. Tremendous progress has been made in finding all irreducible representations for many specific groups.

Representation theory occurs throughout nature. Any time you have a change of coordinate systems, suddenly representations appear. In fact, most theoretical physicists will even define an elementary particle (such as an electron) as an irreducible representation of some group (a group that captures the intrinsic symmetries of the world). For more on this, see Sternberg's *Group Theory and Physics* [106], especially the last part of Chapter 3.9.

11.3 Rings

If groups are roughly viewed as sets for which there is an addition, then rings are sets for which there is both an addition and a multiplication.

Definition 11.3.1 *A nonempty set R is a* ring *if there are two binary operations, denoted by $\cdot$ and $+$, on R such that*

a) *R with $+$ forms an abelian group. The identity is denoted by 0.*

b) *(Associativity) for all $a, b, c \in R$, $a \cdot (b \cdot c) = (a \cdot b) \cdot c$.*

c) *(Distributivity) for all $a, b, c \in R$,*

$$a \cdot (b + c) = a \cdot b + a \cdot c$$

and

$$(a + b) \cdot c = a \cdot c + b \cdot c.$$

Note that rings are not required to be commutative for the $\cdot$ operation or, in other words, we do not require $a \cdot b = b \cdot a$.

If there exists an element $1 \in R$ with $1 \cdot a = a \cdot 1 = a$ for all $a \in R$, we say that R is a ring with unit element. Almost all rings that are ever encountered in life will have a unit element.

The integers $\mathbf{Z} = \{\ldots, -3, -2, -1, 0, 1, 2, 3, \ldots\}$, with the usual addition and multiplication, form a ring. Polynomials in one variable x with complex coefficients, denoted by $\mathbf{C}[x]$, form a ring with the usual addition and multiplication of polynomials. In fact, polynomials in n variables $\{x_1, \ldots, x_n\}$ with complex coefficients, denoted by $\mathbf{C}[x_1, \ldots, x_n]$, will also form a ring in the natural way. By the way, the study of the ring theoretic properties of $\mathbf{C}[x_1, \ldots, x_n]$ is at the heart of much of algebraic geometry. While polynomials with complex coefficients are the most common to study, it is of course the case that polynomials with integer coefficients ($\mathbf{Z}[x_1, \ldots, x_n]$), polynomials with rational coefficients ($\mathbf{Q}[x_1, \ldots, x_n]$) and polynomials with real coefficients ($\mathbf{R}[x_1, \ldots, x_n]$) are also rings. In fact, if R is any ring, then the polynomials with coefficients in R form a ring, denoted by $R[x_1, \ldots, x_n]$.

Definition 11.3.2 *A function* $\sigma : R \to \hat{R}$ *between rings* R *and* $\hat{R}$ *is a* ring homomorphism *if for all* $a, b \in R$,

$$\sigma(a + b) = \sigma(a) + \sigma(b)$$

and

$$\sigma(a \cdot b) = \sigma(a) \cdot \sigma(b).$$

Definition 11.3.3 *A subset* I *of a ring* R *is an* ideal *if* I *is a subgroup of* R *under* $+$ *and if, for any* $a \in R$, $aI \subset I$ *and* $Ia \subset I$.

The notion of an ideal in ring theory corresponds to the notion of a normal subgroup in group theory. This analogy is shown in the following theorems:

Theorem 11.3.1 *Let* $\sigma : R \to \hat{R}$ *be a ring homomorphism. Then the set*

$$ker(\sigma) = \{a \in R : \sigma(a) = 0\}$$

is an ideal in R. *(This ideal* $ker(\sigma)$ *is called the* kernel *of the map* σ.)

Sketch of Proof: We need to use that for all $x \in \hat{R}$,

$$x \cdot 0 = 0 \cdot x = 0,$$

which is an exercise at the end of the chapter. Let $b \in \ker(\sigma)$. Thus $\sigma(b) = 0$. Given any element $a \in R$, we want $a \cdot b \in \ker(\sigma)$ and $b \cdot a \in \ker(\sigma)$. We have

$$\begin{aligned} \sigma(a \cdot b) &= \sigma(a) \cdot \sigma(b) \\ &= \sigma(a) \cdot 0 \\ &= 0, \end{aligned}$$

implying that $a \cdot b \in \ker(\sigma)$.

By a similar argument, $b \cdot a \in \ker(\sigma)$, showing that $\ker(\sigma)$ is indeed an ideal. $\square$

Theorem 11.3.2 *Let* I *be an ideal in* R. *The sets* $\{a + I : a \in R\}$ *form a ring, denoted* R/I, *under the operations* $(a + I) + (b + I) = (a + b + I)$ *and* $(a + I) \cdot (b + I) = (a \cdot b + I)$.

The proof is left as a (long) exercise at the end of the chapter.

The study of a ring comes down to studying its ideals, or equivalently, its homomorphisms. Again, it's a mid-twentieth century approach to translate the study of rings to the study of maps between rings.

11.4 Fields and Galois Theory

We are now ready to enter the heart of classical algebra. To a large extent, the whole point of high school algebra is to find roots of linear and quadratic polynomials. With more complicated, but in spirit, similar techniques, the roots for third and fourth degree polynomials can also be found. One of the main historical motivations for developing the machinery of group and ring theory was in showing that there can be no similar techniques for finding the roots of polynomials of fifth degree or higher. More specifically the roots of a fifth degree or higher polynomial cannot be obtained by a formula involving radicals of the coefficients of the polynomial. (For an historical account, see Edwards' *Galois Theory* [31].)

The key is to establish a correspondence between one variable polynomials and finite groups. This is the essence of Galois Theory, which explicitly connects the ability to express roots as radicals of coefficients (in analogue to the quadratic equation) with properties of the associated group.

Before describing this correspondence, we need to discuss fields and field extensions.

Definition 11.4.1 *A ring R is a* field *if*
1. *R has a multiplicative unit 1,*
2. *for all $a, b \in R$ we have $a \cdot b = b \cdot a$ and*
3. *for any $a \neq 0$ in R, there is an element denoted by a^{-1} with $a \cdot a^{-1} = 1$.*

For example, since the integers $\mathbf{Z}$ do not have multiplicative inverses, $\mathbf{Z}$ is not a field. The rationals $\mathbf{Q}$, the reals $\mathbf{R}$ and the complexes $\mathbf{C}$ are fields. For the ring $\mathbf{C}[x]$ of one variable polynomials, there corresponds the field $\mathbf{C}(x) = \{\frac{P(x)}{Q(x)} : P(x), Q(x) \in \mathbf{C}[x], Q(x) \neq 0\}$.

Definition 11.4.2 *A field $\hat{k}$ is a* field extension *of a field k if k is contained in $\hat{k}$.*

For example, the complex numbers $\mathbf{C}$ is a field extension of the real numbers $\mathbf{R}$.

Once we have the notion of a field, we can form the ring $k[x]$ of one variable polynomials with coefficients in k. Basic, but deep, is:

Theorem 11.4.1 *Let k be a field. Then there is a field extension $\hat{k}$ of k such that every polynomial in $k[x]$ has a root in $\hat{k}$.*

Such a field $\hat{k}$ is said to be *algebraically closed*. For a proof, see Garling's *A Course in Galois Theory* [45], Section 8.2. As a word of warning, the proof uses the Axiom of Choice.

Before showing how groups are related to finding roots of polynomials, recall that the root of a linear equation $ax + b = 0$ is simply $x = -\frac{b}{a}$. For second degree equations, the roots of $ax^2 + bx + c = 0$ are of course

$$x = \frac{-b \pm \sqrt{b^2 - 4ac}}{2a}.$$

Already interesting things are happening. Note that even if the three coefficients a, b and c are real numbers, the roots will be complex if the discriminant $b^2 - 4ac < 0$. Furthermore, even if the coefficients are rational numbers, the roots need not be rational, as $\sqrt{b^2 - 4ac}$ need not be rational.

Both of these observations lead naturally to extension fields of the field of coefficients. We will restrict to the case when the coefficients of our (monic) polynomial are rational numbers.

Let

$$P(x) = x^n + a_{n-1}x^{n-1} + \ldots + a_0,$$

with each $a_k \in \mathbf{Q}$. By the Fundamental Theorem of Algebra (which states that the algebraic closure of the real numbers is the complex numbers), there are complex numbers $\alpha_1, \ldots, \alpha_n$ with

$$P(x) = (x - \alpha)(x - \alpha_2) \cdots (x - \alpha_n).$$

Of course, the whole problem is that the fundamental theorem does not tell us what the roots are. We would like an analogue of the quadratic equation for any degree polynomial. As mentioned before, such analogues do exist for cubic and quartic polynomials, but the punchline of Galois Theory is that no such analogue exists for degree five or higher polynomials. The proof of such a statement involves far more than the tools of high school algebra.

Here is a rapid fire summary of Galois Theory. We will associate to each one variable polynomial with rational coefficients a unique finite dimensional vector space over the rational numbers that is also a field extension of the rational numbers contained in the complex numbers. Namely, if $\alpha_1, \ldots, \alpha_n$ are the roots of the polynomial $P(x)$, the smallest field in the complex numbers that contains both the rationals and the roots $\alpha_1, \ldots, \alpha_n$ is the desired vector space. We then look at all linear transformations from this vector space to itself, with the strong restriction that the linear transformation is also a field automorphism mapping each rational number to itself. This is such a strong restriction that there are only a finite number of such transformations, forming a finite group. Further, each such linear transformation will not only map each root of $P(x)$ to another root but is actually determined by how it maps the roots to each other. Thus the finite group of these special linear transformations are a subgroup of the permutation group on n letters. The final deep result lies in showing that these finite groups determine properties about the roots.

Now for some details. We assume that $P(x)$ is irreducible in $\mathbf{Q}[x]$, meaning that $P(x)$ is not the product of any polynomials in $\mathbf{Q}[x]$. Hence none of the roots α_i of $P(x)$ can be rational numbers.

Definition 11.4.3 *Let $\mathbf{Q}(\alpha_1, \ldots, \alpha_n)$ be the smallest subfield of $\mathbf{C}$ containing both $\mathbf{Q}$ and the roots $\alpha_1, \ldots, \alpha_n$.*

Definition 11.4.4 *Let E be a field extension of $\mathbf{Q}$ but contained in $\mathbf{C}$. We say E is a* splitting field *if there is a polynomial $P(x) \in \mathbf{Q}[x]$ such that $E = \mathbf{Q}(\alpha_1, \ldots, \alpha_n)$, where $\alpha_1, \ldots, \alpha_n$ are the roots in $\mathbf{C}$ of $P(x)$.*

A splitting field E over the rational numbers $\mathbf{Q}$ is in actual fact a vector space over $\mathbf{Q}$. For example, the splitting field $\mathbf{Q}(\sqrt{2})$ is a two-dimensional vector space, since any element can be written uniquely as $a + b\sqrt{2}$, with $a, b \in \mathbf{Q}$.

Definition 11.4.5 *Let E be an extension field of $\mathbf{Q}$. The* group of automorphisms *G of E over $\mathbf{Q}$ is the set of all field automorphisms $\sigma : E \to E$.*

By *field automorphism* we mean a ring homomorphism from the field E to itself that is one-to-one, onto, maps unit to unit and whose inverse is a ring homomorphism. Note that field automorphisms of an extension field have the property that each rational number is mapped to itself (this is an exercise at the end of the chapter).

Such field automorphisms can be interpreted as linear transformations of E to itself. But not all linear transformations are field automorphisms, as will be seen in a moment.

Of course, there is needed here, in a complete treatment, a lemma showing that this set of automorphisms actually forms a group.

Definition 11.4.6 *Given an extension field E over $\mathbf{Q}$ with group of automorphisms G, the* fixed field *of G is the set $\{e \in E : \sigma(e) = e,$ for all $\sigma \in G\}$.*

Note that we are restricting attention to those field automorphisms that contain $\mathbf{Q}$ in the fixed field. Further it can be shown that the fixed field is indeed a subfield of E.

Definition 11.4.7 *A field extension E of* $\mathbf{Q}$ *is normal if the fixed field of the group of automorphisms G of E over* $\mathbf{Q}$ *is exactly* $\mathbf{Q}$.

Let G be the group of automorphisms of $\mathbf{Q}(\alpha_1, \ldots, \alpha_n)$ over $\mathbf{Q}$, where $\mathbf{Q}(\alpha_1, \ldots, \alpha_n)$ is the splitting field of the polynomial

$$
\begin{aligned}
P(x) &= (x - \alpha_1)(x - \alpha_2) \ldots (x - \alpha_n) \\
&= x^n + a_{n-1} x^{n-1} + \ldots + a_0,
\end{aligned}
$$

with each $a_k \in \mathbf{Q}$. This group G is connected to the roots of the polynomial $P(x)$, as seen in:

Theorem 11.4.2 *The group of automorphisms G is a subgroup of the permutation group $\mathbf{S}_n$ on n elements. It is represented by permuting the roots of the polynomial $P(x)$.*

Sketch of Proof: We will show that for any automorphism σ in the group G, the image of every root α_i is another root of $P(x)$. Therefore the automorphisms will merely permute the n roots of $P(x)$. It will be critical that $\sigma(a) = a$ for all rational numbers a. Now

$$
\begin{aligned}
P(\sigma(\alpha_i)) &= (\sigma(\alpha_i))^n + a_{n-1}(\sigma(\alpha_i))^{n-1} + \cdots + a_0 \\
&= \sigma(\alpha_i)^n + \sigma(a_{n-1}(\alpha_i)^{n-1}) + \cdots + \sigma(a_0) \\
&= \sigma((\alpha_i)^n + a_{n-1}(\alpha_i)^{n-1} + \cdots + a_0) \\
&= \sigma(P(\alpha_i)) \\
&= \sigma(0) \\
&= 0.
\end{aligned}
$$

Thus $\sigma(\alpha_i)$ is another root. To finish the proof, which we will not do, we would need to show that an automorphism σ in G is completely determined by its action on the roots α. $\square$

All of this culminates in:

Theorem 11.4.3 (Fundamental Theorem of Galois Theory) *Let $P(x)$ be an irreducible polynomial in $\mathbf{Q}[x]$ and let $E = \mathbf{Q}(\alpha_1, \ldots, \alpha_n)$ be its splitting field with G the automorphism group of E.*

i) Each field B containing $\mathbf{Q}$ *and contained in E is the fixed field of a subgroup of G. Denote this subgroup by G_B.*

ii) The field extension B of $\mathbf{Q}$ *is normal if and only if the subgroup G_B is a normal subgroup of G.*

iii) The rank of E as a vector space over B is the order of G_B. The rank of B as a vector space over $\mathbf{Q}$ *is the order of the group G/G_B.*

Unfortunately, in this brevity, none of the implications should be at all clear. It is not even apparent why this should be called the Fundamental Theorem of the subject. A brief hint or whisper of its importance is that it sets up a dictionary between field extensions B with $\mathbf{Q} \subset \mathbf{B} \subset \mathbf{E}$ and subgroups G_B of G. A see-saw type diagram would be

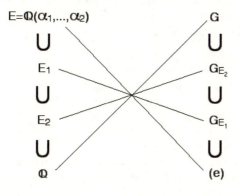

Here the lines connect subgroups with the corresponding fixed fields.

But what does this have to do with finding the roots of a polynomial. Our goal (which Galois Theory shows to be impossible) is to find an analogue of the quadratic equation. We need to make this more precise.

Definition 11.4.8 *A polynomial $P(x)$ is* solvable *if its splitting field $\mathbf{Q}(\alpha_1, \ldots, \alpha_n)$ lies in an extension field of $\mathbf{Q}$ obtained by adding radicals of integers.*

As an example, the field $\mathbf{Q}\{3\sqrt{2}, 5\sqrt{7}\}$ is obtained from $3\sqrt{2}$ and $5\sqrt{7}$, both of which are radicals. On the other hand, the field $\mathbf{Q}(\pi)$ is not obtained by adding radicals to $\mathbf{Q}$; this is a rewording of the deep fact that π is transcendental.

The quadratic equation $x = \frac{-b \pm \sqrt{b^2 - 4ac}}{2a}$ shows that each root of a second degree polynomial can be written in terms of a radical of its coefficients; hence every second degree polynomial is solvable. To show that no analogue of the quadratic equations exists for fifth degree or higher equations, all we need to show is that not all such polynomials are solvable. We want to describe this condition in terms of the polynomial's group of automorphisms.

Definition 11.4.9 *A finite group G is* solvable *if there is a nested sequence of subgroups $G_1, \ldots, G_n$ with $G = G_0 \supseteq G_1 \supseteq G_2 \supseteq \ldots \supseteq G_n = (e)$, with each G_i normal in G_{i-1} and each G_{i-1}/G_i abelian.*

The link between writing roots as radicals and groups is contained in:

Theorem 11.4.4 *A polynomial $P(x)$ is solvable if and only if its associated group G of automorphisms of its splitting field is solvable.*

The impossibility of finding a clean formula for the roots of a high degree polynomial in terms of radicals of the coefficients now follows from showing that generically the group of automorphisms of an nth degree polynomial is the full permutation group $\mathbf{S}_n$ and

Theorem 11.4.5 *The permutation group on n elements, $\mathbf{S}_n$, is not solvable whenever n is greater than or equal to five.*

Of course, these are not obvious theorems. An excellent source for the proofs is Artins' *Galois Theory* [3].

Though there is no algebraic way of finding roots, there are many methods to approximate the roots. This leads to many of the basic techniques in numerical analysis.

11.5   Books

Algebra books went through quite a transformation starting in the 1930s. It was then that Van der Waerden wrote his algebra book *Modern Algebra* [113], which was based on lectures of Emmy Noether. The first undergraduate text mirroring these changes was *A Survey of Modern Algebra* [9], by Garrett Birkhoff and Saunders Mac Lane. The undergraduate text of the sixties and seventies was *Topics in Algebra* by Herstein [57]. Current popular choices are *A First Course in Abstract Algebra* by Fraleigh [41], and *Contemporary Abstract Algebra* by Gallian [43]. Serge Lang's *Algebra* [79] has been for a long time a standard graduate text, though it is not the place to start learning algebra. You will find, in your mathematical career, that you will read many texts by Lang. Jacobson's *Basic Algebra* [68], Artin's *Algebra* [4] and Hungerford's *Algebra* [65] are also good beginning graduate texts.

Galois Theory is definitely one of the most beautiful subjects in mathematics. Luckily there are a number of excellent undergraduate Galois Theory texts. One of the best (and cheapest) is Emil Artin's *Galois Theory* [3]. Other excellent texts are by Ian Stewart [107] and by Garling [45]. Edwards' *Galois Theory* [31] is an historical development. For beginning representation theory, I would recommend Hill's *Groups and Characters* [59] and Sternberg's *Group Theory and Physics* [106].

11.6 Exercises

1. Fix a corner of this book as the origin $(0,0,0)$ in space. Label one of the edges coming out of this corner as the x-axis, one as the y-axis and the last one as the z-axis. The goal of this exercise is to show that rotations do not commute. Let A denote the rotation of the book about the x-axis by ninety degrees and let B be the rotation about the y-axis by ninety degrees. Show with your book and by drawing pictures of your book that applying the rotation A and then rotation B is not the same as applying rotation B first and then rotation A.

2. Prove that the kernel of a group homomorphism is a normal subgroup.

3. Let R be a ring. Show that for all elements x in R,

$$x \cdot 0 = 0 \cdot x = 0,$$

even if the ring R is not commutative.

4. Let R be a ring and I an ideal in the ring. Show that R/I has a ring structure. (This is a long exercise, but it is an excellent way to nail down the basic definition of a ring.)

5. Show that the splitting field $\mathbf{Q}(\sqrt{2})$ over the rational numbers $\mathbf{C}$ is a two dimensional vector space over $\mathbf{C}$.

6. Start with the permutation group $\mathbf{S}_3$.

    a. Find all subgroups of $\mathbf{S}_3$.

    b. Show that the group $\mathbf{S}_3$ is solvable. (This allows us to conclude that for cubic polynomials there is an analogue of the quadratic equation.)

7. For each of the six elements of the group $\mathbf{S}_3$, find the corresponding matrices for the representation of $\mathbf{S}_3$ as described in section two of this chapter.

8. If H is a normal subgroup of a group G, show that there is a natural one-to-one correspondence between the left and the right cosets of H.

9. Let E be a field containing the rational numbers $\mathbf{Q}$. Let σ be a field automorphism of E. Note that this implies in particular that $\sigma(1) = 1$. Show that $\sigma(\frac{p}{q}) = \frac{p}{q}$ for all rational numbers $\frac{p}{q}$.

10. Let $T : G \to \tilde{G}$ be a group homomorphism. Show that $T(g^{-1}) = (T(g))^{-1}$ for all $g \in G$.

11. Let $T : G \to \tilde{G}$ be a group homomorphism. Show that the groups $G/\ker(T)$ and $Im(T)$ are isomorphic. Here $Im(T)$ denotes the image of the group G in the group $\tilde{G}$. This result is usually known as one of the Fundamental Homomorphism Theorems.

Chapter 12

Lebesgue Integration

Basic Object:	Measure Spaces
Basic Map:	Integrable Functions
Basic Goal:	Lebesgue Dominating Convergence Theorem

In calculus we learn about the Riemann integral of a function, which certainly works for many functions. Unfortunately, we must use the word 'many'. Lebesgue measure, and from this the Lebesgue integral, will allow us to define the right notion of integration. Not only will we be able to integrate far more functions with the Lebesgue integral but we will also understand when the integral of a limit of functions is equal to the limit of the integrals, i.e., when

$$\lim_{n \to \infty} \int f_n = \int \lim_{n \to \infty} f_n,$$

which is the Lebesgue Dominating Convergence Theorem. In some sense, the Lebesgue integral is the one that the gods intended us to use all along.

Our approach will be to develop the notion of Lebesgue measure for the real line $\mathbf{R}$, then use this to define the Lebesgue integral.

12.1 Lebesgue Measure

The goal of this section is to define the Lebesgue measure of a set E of real numbers. This intuitively means we want to define the length of E. For intervals

$$E = [a, b] = \{x \in \mathbf{R} : a \le x \le b\}$$

the length of E is simply:

$$\ell(E) = b - a.$$

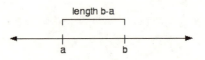

The question is to determine the length of sets that are not intervals, such as

$$E = \{x \in [0,1] : x \text{ is a rational number}\}.$$

We will heavily use that we already know the length of intervals. Let E be any subset of reals. A countable collection of intervals $\{I_n\}$, with each

$$I_n = [a_n, b_n],$$

covers the set E if

$$E \subset \bigcup I_n.$$

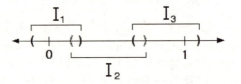

$$E \subset I_1 \cup I_2 \cup I_3$$

Whatever the length or measure of E is, it must be less than the sum of the lengths of the I_n.

Definition 12.1.1 *For any set E in* **R***, the* outer measure *of E is*

$$m^*(E) = \inf\{\sum (b_n - a_n) : \text{The collection of intervals } \{[a_n, b_n]\} \text{ covers } E\}.$$

Definition 12.1.2 *A set E is* measurable *if for every set A,*

$$m^*(A) = m^*(A \cap E) + m^*(A - E).$$

The measure of a measurable set E, denoted by $m(E)$, is $m^*(E)$.

The reason for such a convoluted definition is that not all sets are measurable, though no one will ever construct a nonmeasurable set, since the existence of such a set requires the use of the Axiom of Choice, as we saw in Chapter Ten.

There is another method of defining a measurable set, via the notion of inner measure. Here we define the *inner measure* of a set E to be

$$m_*(E) = \sup(\sum(b_n - a_n) : E \supset \bigcup I_n \text{ and } I_n = [a_n, b_n] \text{ with } a_n \leq b_n\}.$$

Thus instead of covering the set E by a collection of open intervals, we fill up the inside of E with a collection of closed intervals.

If $m^*(E) < \infty$, then the set E can be shown to be measurable if and only if

$$m^*(E) = m_*(E).$$

In either case, we now have a way of measuring the length of almost all subsets of the real numbers.

As an example of how to use these definitions, we will show that the measure of the set of rational numbers (denoted here as E) between 0 and 1 is zero. We will assume that this set E is measurable and show its outer measure is zero. It will be critical that the rationals are countable. In fact, using this countability, list the rationals between zero and one as $a_1, a_2, a_3, \ldots$. Now choose an $\epsilon > 0$. Let I_1 be the interval

$$I_1 = [a_1 - \frac{\epsilon}{2}, a_1 + \frac{\epsilon}{2}].$$

Note that $\ell(I_1) = \epsilon$. Let

$$I_2 = [a_2 - \frac{\epsilon}{4}, a_2 + \frac{\epsilon}{4}].$$

Here $\ell(I_2) = \frac{\epsilon}{2}$. Let

$$I_3 = [a_3 - \frac{\epsilon}{8}, a_3 + \frac{\epsilon}{8}].$$

Here $\ell(I_3) = \frac{\epsilon}{4}$.   In general let

$$I_k = [a_k - \frac{\epsilon}{2^k}, a_k + \frac{\epsilon}{2^k}].$$

Then $\ell(I_k) = \frac{\epsilon}{2^{k-1}}$.

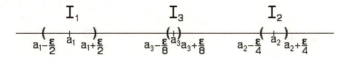

Certainly the rationals between zero and one are covered by this countable collection of open sets :

$$E \subset \bigcup I_n.$$

Then

$$m(E) \leq \sum_{n=1}^{\infty} \ell(I_n)$$

$$= \sum_{k=1}^{\infty} \frac{\epsilon}{2^{k-1}}$$

$$= \epsilon \sum_{k=0}^{\infty} \frac{1}{2^k}$$

$$= 2\epsilon.$$

By letting ϵ approach zero, we see that $m(E) = 0$.

A similar argument can be used to show that the measure of any countable set is zero and in fact appears as an exercise at the end of this chapter.

12.2   The Cantor Set

While long a source of examples and counterexamples in real analysis, the Cantor set has recently been playing a significant role in dynamical systems. It is an uncountable, nowhere dense measure zero subset of the unit interval $[0, 1]$. By nowhere dense, we mean that the closure of the complement of the Cantor set will be the entire unit interval. We will first construct the Cantor set, then show that it is both uncountable and has measure zero.

For each positive integer k, we will construct a subset C_k of the unit interval and then define the Cantor set C to be

$$C = \bigcap_{k=1}^{\infty} C_k.$$

For $k = 1$, split the unit interval $[0, 1]$ into thirds and remove the open middle third, setting

$$C_1 = [0, 1] - (\frac{1}{3}, \frac{2}{3})$$

$$= [0, \frac{1}{3}] \bigcup [\frac{2}{3}, 1].$$

$$0 \qquad \frac{1}{3} \qquad \frac{2}{3} \qquad 1$$

$$\mathbf{C_1}$$

Take these two intervals and split them into thirds. Now remove each of their middle thirds to get

$$C_2 = [0, \frac{1}{9}] \bigcup [\frac{2}{9}, \frac{1}{3}] \bigcup [\frac{2}{3}, \frac{7}{9}] \bigcup [\frac{8}{9}, 1].$$

C_2

To get the next set C_3, split each of the four intervals of C_2 into three equal parts and remove the open middle thirds, to get eight closed intervals, each of length $\frac{1}{27}$. Continue this process for each k, so that each C_k consists of 2^k closed intervals, each of length $\frac{1}{3^k}$. Thus the length of each C_k will be

$$\text{length} = \frac{2^k}{3^k}.$$

The *Cantor set* C is the intersection of all of these C_k:

$$\text{Cantor set} = C = \bigcap_{k=1}^{\infty} C_k.$$

Part of the initial interest in the Cantor set was it was both uncountable and had measure zero. We will show first that the Cantor set has measure zero and then that it is uncountable. Since C is the intersection of all of the C_k, we get for each k that

$$m(C) < m(C_k) = \frac{2^k}{3^k}.$$

Since the fractions $\frac{2^k}{3^k}$ go to zero as k goes to infinity, we see that

$$m(C) = 0.$$

It takes a bit more work to show that the Cantor set is uncountable. The actual proof will come down to applying the trick of Cantor diagonalization, as discussed in Chapter Ten. The first step is to express any real number α in the unit interval $[0, 1]$ in its tri-adic expansion

$$\alpha = \sum_{k=1}^{\infty} \frac{n_k}{3^k},$$

where each n_k is zero, one or two. (This is the three-analog of the decimal expansion $\alpha = \sum_{k=1}^{\infty} \frac{n_k}{10^k}$, where here each $n_k = 0, 1, \ldots, 9$.) We can write the tri-adic expansion in base three notation, to get

$$\alpha = .n_1 n_2 n_3 \ldots$$

As with decimal expansion, the tri-adic expansion's coefficients n_k are unique, provided we always round up. Thus we will always say that

$$.102222\ldots = .11000\ldots$$

The Cantor set C has a particularly clean description in terms of the tri-adic or base three expansions. Namely

$$C = \{.n_1 n_2 n_3 \ldots \mid \text{each } n_k \text{ is either zero or two}\}.$$

Thus the effect of removing the middle thirds from all of the intervals corresponds to allowing no 1's among the coefficients. But then the Cantor set can be viewed as the set of infinite sequences of 0's and 2's, which was shown to be uncountable in the exercises of Chapter Ten.

12.3   Lebesgue Integration

One way to motivate integration is to try to find the area under curves. The Lebesgue integral will allow us to find the areas under some quite strange curves.

By definition the area of a unit square is one.

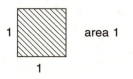

Hence the area of a rectangle with height b and base a will be ab.

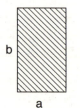

Let E be a measurable set on **R**. Recall that the characteristic function of E, χ_E, is defined by

$$\chi_E(t) = \begin{cases} 1 & \text{if } t \in E \\ 0 & \text{if } t \in \mathbf{R} - E \end{cases}$$

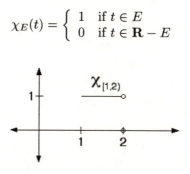

Since the height of χ_E is one, the area under the function (or curve) χ_E must be the length of E, or more precisely, $m(E)$. We denote this by $\int_E \chi_E$. Then the area under the function $a \cdot \chi_E$ must be $a \cdot m(E)$,

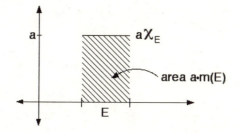

which we denote by $\int_E a\chi_E$.

Now let E and F be disjoint measurable sets. Then the area under the curve $a \cdot \chi_E + b \cdot \chi_F$ must be $a \cdot m(E) + b \cdot m(F)$,

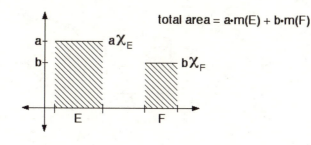

denoted by

$$\int_{E \cup F} a\chi_E + b\chi_F = a \cdot m(E) + b \cdot m(F).$$

For a countable collection of disjoint measurable sets A_i, the function

$$\sum a_i \chi_{Ai}$$

is called a *step function.* Let E be a measurable set. Let

$$\sum a_i \chi_{Ai}$$

be a step function. Then define

$$\int_E (\sum a_i \chi_{Ai}) = \sum a_i m(A_i \cap E).$$

We are about ready to define $\int_E f$.

Definition 12.3.1 *A function* $f : E \to \mathbf{R} \cup (\infty) \cup (-\infty)$ *is measurable if its domain E is measurable and if, for any fixed* $\alpha \in \mathbf{R} \cup (\infty) \cup (-\infty)$,

$$\{x \in E : f(x) = \alpha\}$$

is measurable.

Definition 12.3.2 *Let f be a measurable function on E. Then the* Lebesgue integral *of f on E is*

$$\int_E f = \inf\{\int_E \sum a_i \chi_{A_i} : \text{for all } x \in E, \sum a_i \chi_{A_i}(x) \geq f(x)\}.$$

In pictures:

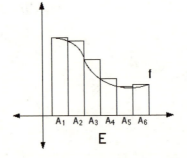

Thus we use that we know the integral for single step functions and then approximate the desired integral by summing the integrals of these step functions.

Every function that is integrable in beginning calculus is Lebesgue integrable. The converse is false, with the canonical counterexample given by the function $f : [0, 1] \to [0, 1]$ which is one at every rational and zero at every irrational. The Lebesgue integral is

$$\int_{[0,1]} f = 0,$$

which is one of the exercises at the end of the chapter, but this function has no Riemann integral, which is an exercise in Chapter Two.

12.4 Convergence Theorems

Not only does the Lebesgue integral allow us to integrate more functions than the calculus class (Riemann) integral, it also provides the right conditions to judge when we can conclude that

$$\int \lim_{k \to \infty} f_k = \lim_{k \to \infty} \int f_k.$$

In fact, if such a result were not true, we would have chosen another definition for the integral.

The typical theorem is of the form:

Theorem 12.4.1 (Lebesgue Dominating Convergence Thm.) *Let $g(x)$ be a Lebesgue integrable function on a measurable set E and let $\{f_n(x)\}$ be a sequence of Lebesgue integrable functions on E with $|f_k(x)| \leq g(x)$ for all x in E and such that there is a pointwise limit of the $f_k(x)$, i.e., there is a function $f(x)$ with*

$$f(x) = \lim_{k \to \infty} f_k(x).$$

Then

$$\int_E \lim_{k \to \infty} f_k(x) = \lim_{k \to \infty} \int_E f_k(x).$$

For a proof, see Royden's *Real Analysis* [95], Chapter 4, in section 4. We will just give a sketch here. Recall that if $f_k(x)$ converges uniformly to $f(x)$, then we know from ϵ and δ real analysis that

$$\lim_{k \to \infty} \int f_k(x) = \int f(x).$$

(i.e., the sequence of functions $f_k(x)$ converges uniformly to $f(x)$ if given any $\epsilon > 0$, there exists a positive integer N with

$$|f(x) - f_k(x)| < \epsilon,$$

for all x and all $k \geq N$. More quaintly, if we put an $\epsilon-tube$ around $y = f(x)$, eventually the $y = f_k(x)$ will fall inside this tube.) The idea in the proof is that the $f_k(x)$ will indeed converge uniformly to $f(x)$, but only away from a subset of E of arbitrarily small measure. More precisely, the proposition we need is:

Proposition 12.4.1 *Let $\{f_n(x)\}$ be a sequence of measurable functions on a Lebesgue measurable set E, with $m(E) < \infty$. Suppose that $\{f_n(x)\}$ converges pointwise to a function $f(x)$. Then given $\epsilon > 0$ and $\delta > 0$, there is a positive integer N and a measurable set $A \subset E$ with $\mid f_k(x) - f(x) \mid < \epsilon$ for all $x \in E - A$ and $k \geq N$ and $m(A) < \delta$.*

The basic idea of the proof of the original theorem is now that

$$\int_E \lim_{n \to \infty} f_n \;=\; \int_{E-A} \lim_{n \to \infty} f_n + \int_A \lim_{n \to \infty} f_n$$

$$=\; \lim_{n \to \infty} \int_{E-A} f_n + max \mid g(x) \mid \cdot m(A).$$

Since we can choose our set A to have arbitrarily small measure, we can let $m(A) \to 0$, which gives us our result.

The proposition can be seen to be true from the following. (After Royden's proof in Chapter 3, Section 6.) Set

$$G_n = \{x \in E : \mid f_n(x) - f(x) \mid \geq \epsilon\}.$$

Set

$$E_N = \bigcup_{n=N}^{\infty} G_n = \{x \in E : \mid f_n(x) - f(x) \mid \geq \epsilon, n \geq N\}.$$

Then $E_{N+1} \subset E_N$. Since we have $f_k(x)$ converging pointwise to $f(x)$, we must have $\cap E_n$, which can be thought of as the limit of the sets E_n, be empty. For measure to have any natural meaning, it should be true that $\lim_{N \to \infty} m(E_N) = 0$. Thus given $\delta > 0$, we can find an E_N with $m(E_N) < \delta$.

This is just an example of what can be accomplished with Lebesgue integration. Historically, the development of the Lebesgue integral in the early part of the twentieth century led quickly to many major advances. For example, until the 1920s, probability theory had no rigorous foundations. With the Lebesgue integral, and thus a correct way of measuring, the foundations were quickly laid.

12.5 Books

One of the first texts on measure theory was by Halmos [54]. This is still an excellent book. The book that I learned measure theory from was Royden's [95] and has been a standard since the 1960s. Rudin's book [96] is another excellent text. Frank Jones, one of the best teachers of mathematics in the country, has recently written a fine text [70]. Folland's recent text [40] is also quite good.

12.6 Exercises

1. Let E be any countable set of real numbers. Show that $m(E) = 0$.

2. Let $f(x)$ and $g(x)$ be two Lebesgue integrable functions, both with domain the set E. Suppose that the set

$$A = \{x \in E : f(x) \neq g(x)\}$$

has measure zero. What can be said about $\int_E f(x)$ and $\int_E g(x)$?

3. Let $f(x) = x$ for all real numbers x between zero and one and let $f(x)$ be zero everywhere else. We know from calculus that

$$\int_0^1 f(x)dx = \frac{1}{2}.$$

Show that this function $f(x)$ is Lebesgue integrable and that its Lebesgue integral is still $\frac{1}{2}$.

4. On the interval $[0, 1]$, define

$$f(x) = \begin{cases} 1 & \text{if } x \text{ is rational} \\ 0 & \text{if } x \text{ is not rational} \end{cases}.$$

Show that $f(x)$ is Lebesgue integrable, with

$$\int_0^1 f(x)dx = 0.$$

Chapter 13

Fourier Analysis

Basic Object:	Real-valued functions with a fixed period
Basic Maps:	Fourier transforms
Basic Goal:	Finding bases for vector spaces of periodic functions

13.1 Waves, Periodic Functions and Trigonometry

Waves occur throughout nature, from water pounding a beach to sound echoing off the walls at a club to the evolution of an electron's state in quantum mechanics. For these reasons, at the least, the mathematics of waves is important. In actual fact, the mathematical tools developed for waves, namely Fourier series (or harmonic analysis), touch on a tremendous number of different fields of mathematics. We will concentrate on only a small sliver and look at the basic definitions, how Hilbert spaces enter the scene, what a Fourier transform looks like and finally how Fourier transforms can be used to help solve differential equations.

Of course, a wave should look like:

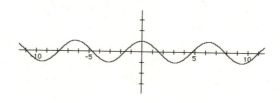

or

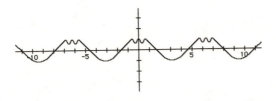

Both of these curves are described by periodic functions.

Definition 13.1.1 *A function $f : \mathbf{R} \to \mathbf{R}$ is* periodic *with* period L *if for all* x, $f(x + L) = f(x)$.

In other words, every L units the function must start to repeat itself. The quintessential periodic functions are the trigonometric functions $\cos(x)$ and $\sin(x)$, each with period 2π. Of course, functions like $\cos(\frac{2\pi x}{L})$ and $\sin(\frac{2\pi x}{L})$ are also periodic, both with period L.

Frequently people will say that a function $f(x)$ has period L if not only do we have that $f(x + L) = f(x)$, but also that there is no smaller number than L for which $f(x)$ is periodic. According to this convention, $\cos(x)$ will have period 2π but not period 4π, despite the fact that, for all x, $\cos(x + 4\pi) = \cos(x)$. We will not follow this convention.

The central result in beginning Fourier series is that almost every periodic function is the, possibly infinite, sum of these trigonometric functions. Thus, at some level, the various functions $\cos(\frac{2\pi x}{L})$ and $\sin(\frac{2\pi x}{L})$ are not merely examples of periodic functions; they generate all periodic functions.

13.2   Fourier Series

Now to see how we can write a periodic function as an (infinite) sum of these cosines and sines. First suppose that we have a function $f : [-\pi, \pi] \to \mathbf{R}$ that has already been written as a series of sines and cosines, namely as

$$a_0 + \sum_{n=1}^{\infty}(a_n \cos(nx) + b_n \sin(nx)).$$

We want to see how we can naively compute the various coefficients a_k and b_k, ignoring all questions of convergence for these infinite series (convergence issues are faced in the next section). For any given k, consider

$$\int_{-\pi}^{\pi} f(x) \cos(kx)\mathrm{d}x = \int_{-\pi}^{\pi}(a_0 + \sum_{n=1}^{\infty}(a_n cos(nx) + b_n sin(nx)))\cos(kx)\mathrm{d}x$$

$$= \int_{-\pi}^{\pi} a_0 \cos(kx)\mathrm{d}x$$

$$+ \sum_{n=1}^{\infty} \int_{-\pi}^{\pi} \cos(nx) \cos(kx)\mathrm{d}x$$

$$+ \sum_{n=1}^{\infty} \int_{-\pi}^{\pi} \sin(nx) \cos(kx)\mathrm{d}x.$$

By direct calculation we have

$$\int_{-\pi}^{\pi} \cos(kx)\mathrm{d}x = \begin{cases} 2\pi & \text{if } k = 0 \\ 0 & \text{if } k \neq 0 \end{cases}$$

$$\int_{-\pi}^{\pi} \cos(nx) \cos(kx)\mathrm{d}x = \begin{cases} \pi & \text{if } k = n \\ 0 & \text{if } k \neq n \end{cases}$$

$$\int_{-\pi}^{\pi} \sin(nx) \cos(kx)\mathrm{d}x = 0.$$

Then we would expect

$$\int_{-\pi}^{\pi} f(x) \cos(kx)\mathrm{d}x = \begin{cases} 2\pi a_0 & \text{if } k = 0 \\ \pi a_n & \text{if } k \neq n \end{cases}.$$

By a similar calculation, using, though, the integrals $\int_{-\pi}^{\pi} f(x) \sin(nx)\mathrm{d}x$, we can get similar formulas for the b_n. This suggests how we could try to write any random periodic function as the infinite sum of sines and cosines:

Definition 13.2.1 *The* Fourier series *for a function* $f : [-\pi, \pi] \to \mathbf{R}$ *is*

$$a_0 + \sum_{n=1}^{\infty} (a_n \cos(nx) + b_n \sin(nx))$$

where

$$a_0 = \frac{1}{2\pi} \int_{-\pi}^{\pi} f(x)\mathrm{d}x$$

and

$$a_n = \frac{1}{\pi} \int_{-\pi}^{\pi} f(x) \cos(nx)\mathrm{d}x$$

and

$$b_n = \frac{1}{\pi} \int_{-\pi}^{\pi} f(x) \sin(nx)\mathrm{d}x.$$

The coefficients a_i *and* b_j *are called the* amplitudes, *or Fourier coefficients for the Fourier series.*

Of course, such a definition can only be applied to those functions for which the above integrals exist. The punchline, as we will see, is that most functions are actually equal to their Fourier series.

There are other ways of writing the Fourier series for a function. For example, using that $e^{ix} = \cos x + i \sin x$, for real numbers x, the Fourier series can also be expressed by

$$\sum_{n=-\infty}^{\infty} C_n e^{inx},$$

where

$$C_n = \frac{1}{2\pi} \int_{-\pi}^{\pi} f(x) e^{inx} \, dx.$$

The C_n are also called the *amplitudes* or *Fourier coefficients*. In fact, for the rest of this section, but not for the rest of the chapter, we will write our Fourier series as $\sum_{n=-\infty}^{\infty} C_n e^{inx}$

The hope (which can be almost achieved) is that the function $f(x)$ and its Fourier series will be equal. For this, we must first put a slight restriction on the type of function we allow.

Theorem 13.2.1 *Let* $f : [-\pi, \pi] \rightarrow \mathbf{R}$ *be a square-integrable function. (i.e.,*

$$\int_{-\pi}^{\pi} |f(x)|^2 dx < \infty.)$$

Then at almost all points,

$$f(x) = \sum_{n=-\infty}^{\infty} C_n e^{inx},$$

its Fourier series.

Note that this theorem contains within it the fact that the Fourier series of a square-integrable function will converge. Further, the above integral is the Lebesgue integral. Recall that almost everywhere means at all points except possibly for points in a set of measure zero. As seen in exercise 2 in Chapter Twelve, two functions that are equal almost everywhere will have equal integrals. Thus, morally, a square-integrable function is equal to its Fourier series.

What the Fourier series does is associate to a function an infinite sequence of numbers, the amplitudes. It explicitly gives how a function is the (infinite) sum of complex waves e^{inx}. Thus there is a map $\Im$ from square-integrable functions to infinite sequences of complex numbers,

$\Im :$	Vector Space of square-integrable functions	$\rightarrow$	Certain vector space of infinite sequences of complex numbers

or

$\Im :$	Vector Space of square-integrable functions	$\rightarrow$	Vector space of infinite sequences of amplitudes

which, by the above theorem, is one-to-one, modulo equivalence of functions almost everywhere.

We now translate these statements into the language of Hilbert spaces, an extremely important class of vector space. Before giving the definition of a Hilbert space, a few definitions must be made.

Definition 13.2.2 *An* inner product $\langle \cdot, \cdot \rangle : V \times V \to \mathbf{C}$ *on a complex vector space* V *is a map such that*

1. $\langle av_1 + bv_2, v_3 \rangle = a \langle v_1, v_3 \rangle + b \langle v_2, v_3 \rangle$ *for all complex numbers* $a, b \in \mathbf{C}$ *and for all vectors* $v_1, v_2, v_3, \in V$.

2. $\langle v, w \rangle = \overline{\langle w, v \rangle}$ *for all* $v, w \in V$.

3. $\langle v, v \rangle \geq 0$ *for all* $v \in V$ *and* $\langle v, v \rangle = 0$ *only if* $v = 0$.

Note that since $\langle v, v \rangle = \overline{\langle v, v \rangle}$, we must have, for all vectors v, that $\langle v, v \rangle$ is a real number. Hence the third requirement that $\langle v, v \rangle \geq 0$ makes sense.

To some extent, this is the complex vector space analogue of the dot product on $\mathbf{R}^n$. In fact, the basic example of an inner product on $\mathbf{C}^n$ is the following: let

$$v = (v_1, \ldots, v_n)$$
$$w = (w_1, \ldots, w_n)$$

be two vectors in $\mathbf{C}^n$. Define

$$\langle v, w \rangle = \sum_{k=1}^{n} v_k \overline{w}_k.$$

It can be checked that this is an inner product on $\mathbf{C}^n$.

Definition 13.2.3 *Given an inner product* $\langle \cdot, \cdot \rangle : V \times V \to \mathbf{C}$, *the* induced norm *on* V *is given by:*

$$|v| = \langle v, v \rangle^{1/2}.$$

In an inner product space, two vectors are *orthogonal* if their inner product is zero (which is what happens for the dot product in $\mathbf{R}^n$). Further, we can interpret the norm of a vector as a measure of the distance from the vector to the origin of the vector space. But then, with a notion of distance, we have a metric and hence a topology on V, as seen in Chapter Four, by setting

$$\rho(v, w) = |v - w|.$$

Definition 13.2.4 *A metric space (X, ρ) is* complete *if every Cauchy sequence converges, meaning that for any sequence $\{v_i\}$ in X with $\rho(v_i, v_j) \to 0$ as $i, j \to \infty$, there is an element v in X with $v_i \to v$ (i.e., $\rho(v, v_i) \to 0$ as $i \to \infty$).*

Definition 13.2.5 *A* Hilbert space *is an inner product space which is complete with respect to the topology defined by the inner product.*

There is the following natural Hilbert space.

Proposition 13.2.1 *The set of Lebesgue square-integrable functions*

$$\mathbf{L}^2[-\pi, \pi] = \{\mathbf{f} : [-\pi, \pi] \to \mathbf{C} \mid \int_{-\pi}^{\pi} |\mathbf{f}|^2 < \infty\}$$

is a Hilbert space, with inner product

$$\langle f, g \rangle = \int_{-\pi}^{\pi} f(x) \cdot \overline{g(x)} \mathrm{d}x.$$

This vector space is denoted by $L^2[-\pi, \pi]$.

We need to allow Lebesgue integrable functions in the above definition in order for the space to be complete.

In general, there is, for each real number $p \geq 1$ and any interval $[a, b]$, the vector space:

$$\mathbf{L}^p[a, b] = \{f : [a, b] \to \mathbf{R} \mid \int_a^b |f(x)|^p \mathrm{d}x < \infty\}.$$

The study of these vector spaces is the start of *Banach Space* theory.

Another standard example of a Hilbert space is the space of square-integrable sequences, denoted by l^2:

Proposition 13.2.2 *The set of sequences of complex numbers*

$$l^2 = \{(a_0, a_1, \ldots) \mid \sum_{j=0}^{\infty} |a_j|^2 < \infty\}$$

is a Hilbert space with inner product

$$\langle (a_0, a_1, \ldots), (b_0, b_1, \ldots) \rangle = \sum_{j=0}^{\infty} a_j \overline{b_j}.$$

We can now restate the fact that square-integrable functions are equal to their Fourier series, almost everywhere, into the language of Hilbert spaces.

Theorem 13.2.2 *For the Hilbert space* $\mathbf{L}^2[-\pi, \pi]$, *the functions*

$$\frac{1}{\sqrt{2\pi}} e^{inx}$$

are an orthonormal (Schauder) basis, meaning that each has length one, that they are pairwise orthogonal and that each element of $\mathbf{L}^2[-\pi, \pi]$ *is the unique infinite linear combination of the basis elements.*

Note that we had to use the technical term of *Schauder* basis. These are not quite the bases defined in Chapter One. There we needed each element in the vector space to be a unique finite linear combination of basis elements. While such do exist for Hilbert spaces, they do not seem to be of much use (the proof of their existence actually stems from the Axiom of Choice). The more natural bases are the above, for which we still require uniqueness of the coefficients but now allow infinite sums.

While the proof that the functions $\frac{1}{\sqrt{2\pi}} e^{inx}$ are orthonormal is simply an integral calculation, the proof that they form a basis is much harder and is in fact a restatement that a square-integrable function is equal to its Fourier series, namely:

Theorem 13.2.3 *For any function* $f(x)$ *in the Hilbert space* $\mathbf{L}^2[-\pi, \pi]$, *we have*

$$f(x) = \sum_{n=-\infty}^{\infty} \langle f(x), \frac{1}{\sqrt{2\pi}} e^{inx} \rangle \frac{1}{\sqrt{2\pi}} e^{inx},$$

almost everywhere.

Hence, the coefficients of a function's Fourier series are simply the inner product of $f(x)$ with each basis vector, exactly as with the dot product for vectors in $\mathbf{R}^3$ with respect to the standard basis $\begin{pmatrix} 1 \\ 0 \\ 0 \end{pmatrix}, \begin{pmatrix} 0 \\ 1 \\ 0 \end{pmatrix}$ and $\begin{pmatrix} 0 \\ 0 \\ 1 \end{pmatrix}$. Further, we can view the association of a function with its Fourier coefficients (with its amplitudes) as a linear transformation

$$\mathbf{L}^2[-\pi, \pi] \to l^2.$$

Naturally enough, these formulas and theorems have versions for functions with period $2L$, when the Fourier series will be:

Definition 13.2.6 *A function $f : [-L, L] \to \mathbf{R}$ has Fourier series*

$$\sum_{n=-\infty}^{\infty} C_n e^{i\frac{n\pi x}{L}},$$

where

$$C_n = \frac{1}{2L} \int_{-L}^{L} f(x) e^{-\frac{in\pi x}{L}} \, dx.$$

We have ignored, so far, a major subtlety, namely that a Fourier series is an infinite series. The next section deals with these issues.

13.3   Convergence Issues

Already during the 1700s mathematicians were trying to see if a given function was equal to its Fourier series, though in actual fact the theoretical tools needed to talk about such questions were not yet available, leading to some nonsensical statements. By the end of the 1800s, building on work of Dirichlet, Riemann and Gibbs, much more was known.

This section will state some of these convergence theorems. The proofs are hard. For notation, let our function be $f(x)$ and denote its Fourier series by

$$a_0 + \sum_{n=1}^{\infty} (a_n \cos(nx) + b_n \sin(nx)).$$

We want to know what this series converges to pointwise and to know when the convergence is uniform.

Theorem 13.3.1 *Let $f(x)$ be continuous and periodic with period 2π. Then*

$$\lim_{N \to \infty} \int_{-\pi}^{\pi} \left(f(x) - [a_0 + \sum_{n=1}^{N} (a_n \cos(nx) + b_n \sin(nx))] \right) dx = 0.$$

Thus for continuous functions, the area under the curve

$$y = \text{partial sum of the Fourier series}$$

will approach the area under the curve $y = f(x)$. We say that the Fourier series *converges in the mean* to the function $f(x)$.

This is telling us little about what the Fourier series converges to at any given fixed point x. Now assume that $f(x)$ is piecewise smooth on the closed interval $[-\pi, \pi]$, meaning that $f(x)$ is piecewise continuous, has a derivative at all but a finite number of points and that the derivative is piecewise continuous. For such functions, we define the one sided limits

$$f(x+) = \lim_{h \to 0 \text{ and } h > 0} f(x + h)$$

and

$$f(x-) = \lim_{h \to 0 \text{ and } h > 0} f(x - h).$$

Theorem 13.3.2 *If $f(x)$ is piecewise smooth on $[-\pi, \pi]$, then for all points x, the Fourier series converges pointwise to the function*

$$\frac{f(x+) + f(x-)}{2}.$$

At points where $f(x)$ is continuous, the one sided limits are of course each equal to $f(x)$. Thus for a continuous, piecewise smooth function, the Fourier series will converge pointwise to the function.

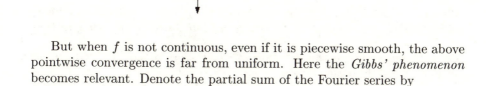

But when f is not continuous, even if it is piecewise smooth, the above pointwise convergence is far from uniform. Here the *Gibbs' phenomenon* becomes relevant. Denote the partial sum of the Fourier series by

$$S_N(x) = \frac{a_0}{2} + \sum_{n=1}^{N} (a_n \cos(nx) + b_n \sin(nx))$$

and suppose that f has a point of discontinuity at x_0. While the partial sums $S_N(x)$ do converge to $\frac{f(x+)+f(x-)}{2}$, the rate of convergence at different x is wildly different. In fact, the better the convergence is at the point of discontinuity x_0, the worse it is near x_0. In pictures, what happens is:

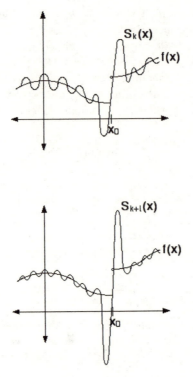

Note how the partial sums soar away from the function $f(x)$, destroying any hope of uniform convergence.

Luckily this does not happen if the function is continuous and piecewise smooth.

Theorem 13.3.3 *Let $f(x)$ be continuous and piecewise smooth on $[-\pi, \pi]$, with $f(-\pi) = f(\pi)$. Then the Fourier series will converge uniformly to $f(x)$.*

Thus for reasonably decent functions, we can safely substitute their Fourier series and still do basic calculus.

For proofs of these results, see Harry F. Davis' *Fourier Series and Orthogonal Functions* [24], chapter 3.

13.4 Fourier Integrals and Transforms

Most functions $f : \mathbf{R} \to \mathbf{R}$ will of course not be periodic, no matter what period L is chosen. But all functions, in some sense, are infinitely periodic. The Fourier integral is the result when we let the period L approach infinity (having as a consequence that $\frac{n\pi x}{L}$ approaches zero). The summation sign in the Fourier series becomes an integral. The result is:

Definition 13.4.1 *Let $f : \mathbf{R} \to \mathbf{R}$ be a function. Its* Fourier integral *is*

$$\int_0^\infty (a(t)\cos(tx) + b(t)\sin(tx))\mathrm{d}t,$$

where

$$a(t) = \frac{1}{\pi} \int_{-\infty}^\infty f(x)\cos(tx)\mathrm{d}x$$

and

$$b(t) = \frac{1}{\pi} \int_{-\infty}^\infty f(x)\sin(tx)\mathrm{d}x.$$

The Fourier integral can be rewritten as

$$\int_{-\infty}^\infty C(t)e^{itx}\mathrm{d}t,$$

where

$$C(t) = \frac{1}{2\pi} \int_{-\infty}^\infty f(x)e^{itx}\mathrm{d}x.$$

There are other forms, all equivalent up to constants.

The main theorem is:

Theorem 13.4.1 *Let $f : \mathbf{R} \to \mathbf{R}$ be integrable (i.e., $\int_{-\infty}^\infty |f(x)|\mathrm{d}x < \infty$). Then, off of a set of measure zero, the function $f(x)$ is equal to its Fourier integral.*

As with Fourier series, this integral is the Lebesgue integral. Further, again recall that by the term 'a set of measure zero', we mean a set of Lebesgue measure zero and that throughout analysis, sets of measure zero are routinely ignored.

As we will see, a large part of the usefulness of Fourier integrals lies in the existence of the Fourier transform.

Definition 13.4.2 *The* Fourier transform *of an integrable function $f(x)$ is:*

$$\Im(f(x))(t) = \int_{-\infty}^\infty f(x)e^{-itx}\mathrm{d}x.$$

The idea is that the Fourier transform can be viewed as corresponding to the coefficients a_n and b_n of a Fourier series and hence to the amplitude of the wave. By a calculation, we see that

$$f(x) = \frac{1}{2\pi} \int_{-\infty}^{\infty} \Im(f(x))(t)e^{itx} \, \mathrm{d}t,$$

provided we place suitable restrictions on the function $f(x)$. Thus indeed the Fourier transform is the continuous analogue of the amplitudes for Fourier series, in that we are writing the original function $f(x)$ as a sum (an integral) of the complex waves e^{itx} with coefficients given by the transform. (Also, the constant $\frac{1}{2\pi}$ is not fixed in stone; what is required is that the product of the constants in front of the integral in the Fourier transform (here it is 1) and the above integral be equal to $\frac{1}{2\pi}$.)

As we will see in the next section, in applications you frequently know the Fourier transform before you know the original function.

But for now we can view the Fourier transform as a one-to-one map

$$\Im : \text{Vector Space of Functions} \to \text{Different Vector Space of Functions.}$$

Thinking of the Fourier transform as an amplitude, we can rewrite this as:

$$\Im : \text{Position Space} \to \text{Amplitude Space.}$$

Following directly from the linearity of the Lebesgue integral, this map is linear.

Much of the power of Fourier transforms is that there is a dictionary between the algebraic and analytic properties of the functions in one of these vector spaces with those of the other vector space.

Proposition 13.4.1 *Let $f(x,t)$ be an integrable function with $f(x,t) \to 0$ as $x \to \pm\infty$. Let $\Im(f(x))(u)$ denote the Fourier transform with respect to the variable x. Then*

i) $\Im\{\frac{\partial f}{\partial x}\}(u) = iu\Im(f(x))(u).$

ii) $\Im\{\frac{\partial^2 f}{\partial x^2}\}(u) = -u^2\Im(f(x))(u).$

iii) $\Im\{\frac{\partial f(x,t)}{\partial t}\}(u) = \frac{\partial}{\partial t}\{\Im(f(x,t))\}(u).$

We will show (i), where the key tool is simply integration by parts and sketch the proof of (iii).

By the definition of the Fourier transform, we have

$$\Im\{\frac{\partial f}{\partial x}\}(u) = \int_{-\infty}^{\infty} \frac{\partial f}{\partial x}e^{-iux} \, \mathrm{d}x,$$

which, by integration by parts, is

$$e^{-iux} f(x,t) \mid_{-\infty}^{\infty} +iu \int_{-\infty}^{\infty} f(x,t) e^{-iux} \mathrm{d}x = iu \int_{-\infty}^{\infty} f(x,t) e^{iux} \mathrm{d}x,$$

since $f(x,t) \to 0$ as $x \to \pm\infty$, and hence equals

$$iu\Im(f).$$

For (iii), we have

$$\Im\{\frac{\partial f(x,t)}{\partial t}\}(u) = \int_{-\infty}^{\infty} \frac{\partial f(x,t)}{\partial t} e^{-iux} \mathrm{d}x.$$

Since this integral is with respect to x and since the partial derivative is with respect to t, this is equal to:

$$\frac{\partial}{\partial t} \int_{-\infty}^{\infty} f(x,t) e^{-iux} \mathrm{d}x.$$

But this is just:

$$\frac{\partial}{\partial t}\{\Im(f(x,t))\}(u),$$

and thus (iii) has been shown. $\square$

In the next section we will use this proposition to reduce the solving of a partial differential equation to the solving of an ordinary differential equation (which can almost always be solved). We need one more preliminary definition.

Definition 13.4.3 *The convolution of two functions $f(x)$ and $g(x)$ is*

$$(f * g)(x) = \int_{-\infty}^{\infty} f(u)g(x-u)\mathrm{d}u.$$

By a direct calculation, the Fourier transform of a convolution is the product of the Fourier transforms of each function, i.e.,

$$\Im(f * g) = \Im(f) \cdot \Im(g).$$

Thus the Fourier transform translates a convolution in the original vector space into a product in the image vector space. This will be important when trying to solve partial differential equations, in that at some stage we will have the product of two Fourier transforms, which we can now recognize as the Fourier transform of a single function, the convolution.

13.5   Solving Differential Equations

The idea is that the Fourier transform will translate a differential equation
into a simpler one (one that is, vaguely, more algebraic). We will apply
this technique to solving the partial differential equation that describes the
flow of heat. Here the Fourier transform will change the partial differential
equation into an ordinary differential equation, which can be solved. Once
we know the Fourier transform, we can almost always recover the original
function.

In the next chapter, we will derive the heat equation, but for now we
will take as a given that the flow of heat through an infinitely thin, long
bar is described by

$$\frac{\partial h}{\partial t} = c\frac{\partial^2 h}{\partial x^2},$$

where $h(x,t)$ denotes the temperature at time t and position x and where c
is a given constant. We start with an initial temperature distribution $f(x)$.
Thus we want to find a function $h(x,t)$ that satisfies

$$\frac{\partial h}{\partial t} = c\frac{\partial^2 h}{\partial x^2},$$

given the initial condition,

$$h(x,0) = f(x).$$

Further, assume that as $x \to \pm\infty$, we know that $f(x) \to 0$. This just
means basically that the bar will initially have zero temperature for large
values of x. For physical reasons we assume that whatever is the eventual
solution $h(x,t)$, we have that $h(x,t) \to 0$ as $x \to \pm\infty$.

Take the Fourier transform with respect to the variable x of the partial
differential equation

$$\frac{\partial h}{\partial t} = k \cdot \frac{\partial^2 h}{\partial x^2},$$

to get

$$\Im(\frac{\partial h(x,t)}{\partial t})(u) = \Im(k \cdot \frac{\partial^2 h(x,t)}{\partial x^2})(u),$$

yielding

$$\frac{\partial}{\partial t}\Im(h(x,t))(u) = -ku^2\Im(h(x,t))(u).$$

Now $\Im(h(x,t))(u)$ is a function of the variables u and t. The x is a mere symbol, a ghost reminding us of the original PDE.

Treat the variable u as a constant, which is of course what we are doing when we take the partial derivative with respect to t. Then we can write the above equation in the form of an ODE:

$$\frac{\mathrm{d}}{\mathrm{d}t}\Im(h(x,t))(u) = -ku^2\Im(h(x,t))(u).$$

The solution to this ODE, as will be discussed in the next section but which can also be seen directly by (unpleasant) inspection, is:

$$\Im(h(x,t))(u) = C(u)e^{-ku^2t},$$

where $C(u)$ is a function of the variable u alone and hence, as far as the variable t is concerned, is a constant. We will first find this $C(u)$ by using the initial temperature $f(x)$. We know that $h(x,0) = f(x)$. Then for $t = 0$,

$$\Im(h(x,0))(u) = \Im(f(x))(u).$$

When $t = 0$, the function $C(u)e^{-ku^2t}$ is just $C(u)$ alone. Thus when $t = 0$, we have

$$\Im(f(x))(u) = C(u).$$

Since $f(x)$ is assumed to be known, we can actually compute its Fourier transform and thus we can compute $C(u)$. Thus

$$\Im(h(x,t))(u) = \Im(f(x))(u) \cdot e^{-ku^2t}.$$

Assume for a moment that we know a function $g(x,t)$ such that its Fourier transform with respect to x is:

$$\Im(g(x,t))(u) = e^{-ku^2t}.$$

If such a function $g(x,t)$ exists, then

$$\Im(h(x,t))(u) = \Im(f(x))(u) \cdot \Im(g(x,t))(u).$$

But a product of two Fourier transforms can be written as the Fourier transform of a convolution. Thus

$$\Im(h(x,t))(u) = \Im(f(x) * g(x,t)).$$

Since we can recover that original function from its Fourier transform, this means that the solution to the heat equation is

$$h(x,t) = f(x) * g(x,t).$$

Thus we can solve the heat equation if we can find this function $g(x,t)$ whose Fourier transform is e^{-ku^2t}. Luckily we are not the first people to attempt this approach. Over the years many such calculations have been done and tables have been prepared, listing such functions. (To do it oneself, one needs to define the notion of the inverse Fourier transform and then to take the inverse Fourier transform of the function e^{-ku^2t}; while no harder than the Fourier transform, we will not do it.) However it is done, we can figure out that

$$\Im(\frac{1}{\sqrt{4\pi kt}}e^{\frac{-x^2}{4kt}}) = e^{-ku^2t}.$$

Thus the solution of the heat equation will be:

$$h(x,t) = f(x) * \frac{1}{\sqrt{4\pi kt}}e^{\frac{-x^2}{4kt}}.$$

13.6   Books

Since Fourier analysis has applications ranging from CAT scans to questions about the distribution of the prime numbers, it is not surprising that there are books on Fourier series aimed at wildly different audiences and levels of mathematical maturity. Barbara Hubbard's *The World According to Wavelets* [63] is excellent. The first half is a gripping nontechnical description of Fourier series. The second half deals with the rigorous mathematics. Wavelets, by the way, are a recent innovation in Fourier series that have had profound practical applications. A solid, traditional introduction is given by Davis in his *Fourier Series and Orthogonal Functions* [24]. A slightly more advanced text is Folland's *Fourier Analysis and its Applications* [38]. A brief, interesting book is Seeley's *An Introduction to Fourier Series and Integrals* [98]. An old fashioned but readable book is Jackson's *Fourier Series and Orthogonal Polynomials* [67]. For the hardcore student, the classic inspiration in the subject since the 1930s has been Zygmund's *Trigonometric Series* [116].

13.7   Exercises

1. On the vector space

$$\mathbf{L}^2[-\pi,\pi] = \{f : [-\pi,\pi] \to \mathbf{C} \mid \int_{-\pi}^{\pi} |f|^2 < \infty\},$$

show that

$$\langle f, g \rangle = \int_{-\pi}^{\pi} f(x) \cdot \overline{g(x)} \mathrm{d}x$$

is indeed an inner product, as claimed in this chapter.

2. Using Fourier transforms, reduce the solution of the wave equation

$$\frac{\partial^2 y}{\partial t^2} = k \frac{\partial^2 y}{\partial x^2},$$

with k a constant, to solving an ordinary (no partial derivatives involved) differential equation.

3. Consider the functions

$$f_n(x) = \begin{cases} 2n & \text{if } \frac{-1}{n} < x < \frac{1}{n} \\ 0 & \text{otherwise} \end{cases}.$$

Compute the Fourier transforms of each of the functions $f_n(x)$. Graph each of the functions f_n and each of the Fourier transforms. Compare the graphs and draw conclusions.

Chapter 14

Differential Equations

Basic Object:	Differential Equations
Basic Goal:	Finding Solutions to Differential Equations

14.1 Basics

A differential equation is simply an equation, or a set of equations, whose unknowns are functions which must satisfy (or solve) an equation involving both the function and its derivatives. Thus

$$\frac{\mathrm{d}y}{\mathrm{d}x} = 3y$$

is a differential equation whose unknown is the function $y(x)$. Likewise,

$$\frac{\partial^2 y}{\partial x^2} - \frac{\partial^2 y}{\partial x \partial t} + \frac{\partial y}{\partial x} = x^3 + 3yt$$

is a differential equation with the unknown being the function of two variables $y(x, t)$. Differential equations fall into two broad classes: ordinary and partial. Ordinary differential equations (ODEs) are those for which the unknown functions are functions of only one independent variable. Thus $\frac{\mathrm{d}y}{\mathrm{d}x} = 3y$ and

$$\frac{\mathrm{d}^2 y}{\mathrm{d}x^2} + \frac{\mathrm{d}y}{\mathrm{d}x} + \sin(x)y = 0$$

are both ordinary differential equations. As will be seen in the next section, these almost always have, in principle, solutions.

Partial differential equations (PDEs) have unknowns that are functions of more than one variable, such as

$$\frac{\partial^2 y}{\partial x^2} - \frac{\partial^2 y}{\partial t^2} = 0$$

and

$$\frac{\partial^2 y}{\partial x^2} + \left(\frac{\partial y}{\partial t}\right)^3 = \cos(xt).$$

Here the unknown is the function of two variables $y(x, t)$. For PDEs, everything is much murkier as far as solutions go. We will discuss the method of separation of variables and the method of clever change of variables (if this can be even called a method). A third method, discussed in Chapter Thirteen, is to use Fourier transforms.

There is another broad split in differential equations: linear and nonlinear. A differential equation is *homogeneous linear* if given two solutions f_1 and f_2 and any two numbers λ_1 and λ_2, then the function

$$\lambda_1 f_1 + \lambda_2 f_2$$

is another solution. Thus the solutions will form a vector space. For example, $\frac{\partial^2 y}{\partial x^2} - \frac{\partial^2 y}{\partial t^2} = 0$ is homogeneous linear. The differential equation is *linear* if by subtracting off from the differential equation a function of the independent variables alone changes it into a homogeneous linear differential equation. The equation $\frac{\partial^2 y}{\partial x^2} - \frac{\partial^2 y}{\partial t^2} = x$ is linear, since if we subtract off the function x we have a homogeneous linear equation. The important fact about linear differential equations is that their solution spaces form linear subspaces of vector spaces, allowing linear algebraic ideas to be applied. Naturally enough a *nonlinear* differential equation is one which is not linear.

In practice, one expects to have differential equations arise whenever one quantity varies with respect to another. Certainly the basic laws of physics are written in terms of differential equations. After all, Newton's second law:

$$\text{Force} = (\text{mass}) \cdot (\text{acceleration})$$

is the differential equation

$$\text{Force} = (\text{mass}) \cdot \left(\frac{\mathrm{d}^2(position)}{\mathrm{d}x^2}\right).$$

14.2   Ordinary Differential Equations

In solving an ordinary differential equation, one must basically undo a derivative. Hence solving an ordinary differential equation is basically the same as performing an integral. In fact, the same types of problems occur in ODEs and in integration theory.

Most reasonable functions (such as continuous functions) can be integrated. But to actually recognize the integral of a function as some other, well-known function (such as a polynomial, trig function, inverse trig function, exponential or log) is usually not possible. Likewise with ODEs, while almost all have solutions, only a handful can be solved cleanly and explicitly. Hence the standard sophomore-level engineering-type ODE course must inherently have the feel of a bag of tricks applied to special equations.[1]

In this section we are concerned with the fact that ODEs have solutions and that, subject to natural initial conditions, the solutions will be unique. We first see how the solution to a single ODE can be reduced to solving a system of first order ODEs, which are equations with unknown functions $y_1(x), \ldots, y_n(x)$ satisfying

$$\frac{\mathrm{d}y_1}{\mathrm{d}x} = f_1(x, y_1, \ldots, y_n)$$

$$\vdots$$

$$\frac{\mathrm{d}y_n}{\mathrm{d}x} = f_n(x, y_1, \ldots, y_n)$$

Start with a differential equation of the form:

$$a_n(x)\frac{\mathrm{d}^n y}{\mathrm{d}x^n} + \ldots + a_1(x)\frac{\mathrm{d}y}{\mathrm{d}x} + a_0(x)y(x) + b(x) = 0.$$

We introduce new variables:

$$y_0(x) = y(x)$$
$$y_1(x) = \frac{\mathrm{d}y_0}{\mathrm{d}x} = \frac{\mathrm{d}y}{\mathrm{d}x}$$
$$y_2(x) = \frac{\mathrm{d}y_1}{\mathrm{d}x} = \frac{\mathrm{d}^2 y_0}{\mathrm{d}x^2} = \frac{\mathrm{d}^2 y}{\mathrm{d}x^2}$$
$$y_{n-1}(x) = \frac{\mathrm{d}y_{n-2}}{\mathrm{d}x} = \ldots = \frac{\mathrm{d}^{n-1} y_0}{\mathrm{d}x^{n-1}} = \frac{\mathrm{d}^{n-1} y}{\mathrm{d}x^{n-1}}.$$

Then a solution $y(x)$ to the original ODE will give rise to a solution of the following system of first order ODEs:

$$\frac{\mathrm{d}y_0}{\mathrm{d}x} = y_1$$
$$\frac{\mathrm{d}y_1}{\mathrm{d}x} = y_2$$

$$\vdots$$

---

[1]There are reasons and patterns structuring the bag of tricks. These involve a careful study of the underlying symmetries of the equations. For more, see Peter Olver's *Applications of Lie Groups to Differential Equations* [90].

$$\frac{\mathrm{d}y_{n-1}}{\mathrm{d}x} = -\frac{1}{a_n(x)}(a_{n-1}(x)y_{n-1} + a_{n-2}(x)y_{n-2} + \ldots + a_0(x)y_0 + b(x)).$$

If we can solve all such systems of first order ODEs, we can then solve all ODEs. Hence the existence and uniqueness theorems for ODEs can be couched in the language of systems of first order ODEs.

First to define the special class of functions we are interested in.

Definition 14.2.1 *A function* $f(x, y_1, \ldots, y_n)$ *defined on a region* T *in* $\mathbf{R}^{n+1}$ *is Lipschitz if it is continuous and if there is a constant* N *such that for every* $(x, y_1, \ldots, y_n)$ *and* $(\hat{x}, \hat{y_1}, \ldots, \hat{y_n})$ *in* T, *we have*

$$|f(x, y_1, \ldots, y_n) - f(\hat{x}, \hat{y_1}, \ldots, \hat{y_n})| \leq N \cdot (|y_1 - \hat{y_1}| + \ldots + |y_n - \hat{y_n}|).$$

It is not a major restriction on a function to require it to be Lipschitz. For example, any function with continuous first partial derivatives on an open set will be Lipschitz on any connected compact subset.

Theorem 14.2.1 *A system of first order ordinary differential equations*

$$\frac{\mathrm{d}y_1}{\mathrm{d}x} = f_1(x, y_1, \ldots, y_n)$$

$$\vdots$$

$$\frac{\mathrm{d}y_n}{\mathrm{d}x} = f_n(x, y_1, \ldots, y_n),$$

with each function $f_1, \ldots, f_n$ *being Lipschitz in a region* T, *will have, for each real number* x_0, *an interval* $(x_0 - \epsilon, x_0 + \epsilon)$ *on which there are solutions* $y_1(x), \ldots, y_n(x)$. *Further, given numbers* $a_1, \ldots, a_n$, *with* $(x_0, a_1, \ldots, a_n)$ *in the region* T, *the solutions satisfying the initial conditions*

$$y_1(x_0) = a_1$$

$$\vdots$$

$$y_n(x_0) = a_n$$

are unique.

Consider a system of two first order ODEs:

$$\frac{\mathrm{d}y_1}{\mathrm{d}x} = f_1(x, y_1, y_2)$$

$$\frac{\mathrm{d}y_2}{\mathrm{d}x} = f_2(x, y_1, y_2).$$

Then a solution $(y_1(x), y_2(x))$ will be a curve in the plane $\mathbf{R}^2$. The theorem states that there is exactly one solution curve passing through any given point (a_1, a_2). In some sense the reason why ODEs are easier to solve than PDEs is that we are trying to find solution curves for ODEs (a one-dimensional type problem) while for PDEs the solution sets will have higher dimensions and hence far more complicated geometries.

We will set up the *Picard Iteration* for finding solutions and then briefly describe why this iteration actually works in solving the differential equations.

For this iterative process, functions $y_{1_k}(x), \ldots, y_{n_k}(x)$ will be constructed that will approach the true solutions $y_1(x), \ldots, y_n(x)$. Start with setting

$$y_{i_0}(x) = a_i$$

for each i. Then, at the kth step, define

$$y_{1_k}(x) = a_1 + \int_{x_0}^x f_1(t, y_{1_{k-1}}(t), \ldots, y_{n_{k-1}}(t)) dt$$

$$\vdots$$

$$y_{n_k}(x) = a_n + \int_{x_0}^x f_n(t, y_{1_{k-1}}(t), \ldots, y_{n_{k-1}}(t)) dt.$$

The crucial part of the theorem is that each of these converges to a solution. The method is to look at the sequence, for each i,

$$y_{i_0}(x) + \sum_{k=1}^{\infty} (y_{i_k}(x) - y_{i_{k-1}}(x)),$$

which has as its Nth partial sum the function $y_{i_N}(x)$. To show that this sequence converges comes down to showing that

$$|y_{i_k}(x) - y_{i_{k-1}}(x)|$$

approaches zero quickly enough. But this absolute value is equal to

$$\left| \int_{x_0}^x [f_i(t, y_{1_{k-1}}(t), \ldots, y_{n_{k-1}}(t)) - f_i(t, y_{1_{k-2}}(t), \ldots, y_{n_{k-2}}(t))] dt \right|$$

$$\leq \int_{x_0}^x |f_i(t, y_{1_{k-1}}(t), \ldots, y_{n_{k-1}}(t)) - f_i(t, y_{1_{k-2}}(t), \ldots, y_{n_{k-2}}(t))| dt.$$

The last integral's size can be controlled by applying the Lipschitz conditions and showing that it approaches zero.

14.3   The Laplacian

14.3.1   Mean Value Principle

In $\mathbf{R}^n$, the *Laplacian* of a function $u(x) = u(x_1, \ldots, x_n)$ is

$$\triangle u = \frac{\partial^2 u}{\partial x_1{}^2} + \ldots + \frac{\partial^2 u}{\partial x_n{}^2}.$$

One can check that the PDE

$$\triangle u = 0$$

is homogeneous and linear and thus that the solutions form a vector space. These solutions are important enough to justify their own name.

Definition 14.3.1 *A function* $u(x) = u(x_1, \ldots, x_n)$ *is* harmonic *if* $u(x)$ *is a solution to the Laplacian:*

$$\triangle u = 0.$$

Much of the importance of the Laplacian is that its solutions, harmonic functions, satisfy the Mean Value Principle, which is our next topic. For any point $a \in \mathbf{R}^n$, let

$$S_a(r) = \{x \in \mathbf{R}^n : |x - a| = r\},$$

be the sphere of radius r centered at a.

Theorem 14.3.1 (Mean Value Principle) *If* $u(x) = u(x_1, \ldots, x_n)$ *is harmonic, then at any point* $a \in \mathbf{R}^n$,

$$u(a) = \frac{1}{area\ of\ S_a(r)} \int_{S_a(r)} u(x).$$

Thus $u(a)$ is equal to the average value of $u(x)$ on any sphere centered at a. For a proof of the case when n is two, see almost any text on complex analysis. For the general case, see G. Folland's *Introduction to Partial Differential Equations* [39], section 2.A.

Frequently, in practice, people want to find harmonic functions on regions subject to given boundary conditions. This is called:

The Dirichlet Problem: Let R be a region in $\mathbf{R}^n$ with boundary ∂R. Suppose that g is a function defined on this boundary. The *Dirichlet Problem* is to find a function f on R satisfying

$$\triangle f = 0$$

on R and

$$f = g$$

on ∂R.

One way this type of PDE arises naturally in classical physics is as a potential. It is also the PDE used to study a steady-state solution of the heat equation. We will see in the next section that heat flow satisfies the PDE:

$$\frac{\partial^2 u}{\partial x_1^2} + \ldots + \frac{\partial^2 u}{\partial x_n^2} = c \cdot \frac{\partial u}{\partial t},$$

where $u(x_1, \ldots, x_n, t)$ denotes the temperature at time t at place $(x_1, \ldots, x_n)$. By a steady-state solution, we mean a solution that does not change over time, hence a solution with

$$\frac{\partial u}{\partial t} = 0.$$

Thus a steady state solution will satisfy

$$\triangle u = \frac{\partial^2 u}{\partial x_1^2} + \ldots + \frac{\partial^2 u}{\partial x_n^2} = 0,$$

and hence is a harmonic function.

14.3.2 Separation of Variables

There are a number of ways of finding harmonic functions and of solving the Dirichlet Problem, at least when the involved regions are reasonable. Here we discuss the method of separation of variables, a method that can also frequently be used to solve the heat equation and the wave equation. By the way, this technique does not always work.

We will look at a specific example and try to find the solution function $u(x, y)$ to

$$\frac{\partial^2 u}{\partial x^2} + \frac{\partial^2 u}{\partial y^2} = 0,$$

on the unit square, with boundary conditions

$$u(x, y) = \begin{cases} h(x) & \text{if } y = 1 \\ 0 & \text{if } x = 0, \ x = 1 \text{ or } y = 0 \end{cases}$$

where $h(x)$ is some initially specified function defined on the top side of the square.

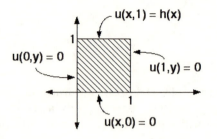

The key assumption will be that the solution will be of the form

$$u(x,y) = f(x) \cdot g(y),$$

where

$$f(0) = 0, \ g(0) = 0, \ f(1) = 0, \ f(x) \cdot g(1) = h(x).$$

This is wild. Few two-variable functions can be written as the product of two functions, each a function of one-variable alone. The only possible justification is if we can actually find such a solution, which is precisely what we will do. (To finish the story, which we will not do, we would need to prove that this solution is unique.) If $u(x,y) = f(x) \cdot g(y)$ and if $\triangle u = 0$, then we need

$$\frac{\mathrm{d}^2 f}{\mathrm{d}x^2} g(y) + f(x)\frac{\mathrm{d}^2 g}{\mathrm{d}y^2} = 0.$$

Thus we would need

$$\frac{\frac{\mathrm{d}^2 f}{\mathrm{d}x^2}}{f(x)} = -\frac{\frac{\mathrm{d}^2 g}{\mathrm{d}y^2}}{g(y)}.$$

Each side depends on totally different variables, hence each must equal to a constant. Using the boundary conditions $f(0) = f(1) = 0$, one can show that this constant must be negative. We denote it by $-c^2$. Thus we need

$$\frac{\mathrm{d}^2 f}{\mathrm{d}x^2} = -c^2 f(x)$$

and

$$\frac{\mathrm{d}^2 g}{\mathrm{d}y^2} = c^2 g(y),$$

both second order ODEs, which have solutions

$$f(x) = \lambda_1 \cos(cx) + \lambda_2 \sin(cx)$$

and

$$g(y) = \mu_1 e^{cy} + \mu_2 e^{-cy}.$$

We now apply the boundary conditions. We have that $f(0) = 0$, which implies that

$$\lambda_1 = 0.$$

Also $g(0) = 0$ forces

$$\mu_1 = -\mu_2$$

and $f(1) = 0$ means that

$$\lambda_2 \sin(cx) = 0.$$

This condition means that the constant c must be of the form

$$c = k\pi, \text{ with } k = 0, 1, 2, \ldots.$$

Hence the solution must have the form

$$u(x, y) = f(x) \cdot g(y) = C_k \sin(k\pi x)(e^{k\pi y} - e^{-k\pi y}),$$

with C_k some constant.

But we also want $u(x, 1) = h(x)$. Here we need to use that the Laplacian is linear and thus that solutions can be added. By adding our various solutions for particular $c = k\pi$, we set

$$u(x, y) = \sum C_k (e^{k\pi y} - e^{-k\pi y}) \sin(k\pi x).$$

All that is left is to find the constants C_k. Since we require $u(x, 1) = h(x)$, we must have

$$h(x) = \sum C_k (e^{k\pi} - e^{-k\pi}) sin(k\pi x).$$

But this is a series of sines. By the Fourier analysis developed in the last chapter, we know that

$$C_k(e^{k\pi} - e^{-k\pi}) = 2 \int_0^1 h(x) \sin(k\pi x) dx = \frac{2h(x)(1 - \cos k\pi)}{k\pi}.$$

Thus the solution is

$$u(x, y) = \frac{2h(x)}{\pi} \sum_{k=1}^{\infty} \frac{1 - \cos k\pi}{k(e^{k\pi} - e^{-k\pi})} \sin(k\pi x)(e^{k\pi y} - e^{-k\pi y}).$$

While not pleasant looking, it is an exact solution.

14.3.3   Applications to Complex Analysis

We will now quickly look at an application of harmonic functions. The goal of Chapter Nine was the study of complex analytic functions $f : U \to \mathbf{C}$, where U is an open set in the complex numbers. One method of describing such $f = u + iv$ was that the real and imaginary parts of f had to satisfy the Cauchy-Riemann equations:

$$\frac{\partial u(x,y)}{\partial x} = \frac{\partial v(x,y)}{\partial y}$$

and

$$\frac{\partial u(x,y)}{\partial y} = -\frac{\partial v(x,y)}{\partial x}.$$

Both real-valued functions u and v are harmonic. The harmonicity of u (and in a similar fashion that of v) can be seen, using the Cauchy-Riemann equations, via:

$$
\begin{aligned}
\Delta u &= \frac{\partial^2 u}{\partial x^2} + \frac{\partial^2 u}{\partial y^2} \\
&= \frac{\partial}{\partial x}\frac{\partial v}{\partial y} + \frac{\partial}{\partial y}\frac{-\partial v}{\partial x} \\
&= 0.
\end{aligned}
$$

One approach to complex analysis is to push hard on the harmonicity of the real-valued functions u and v.

14.4   The Heat Equation

We will first describe the partial differential equation that is called the Heat Equation and then give a physics-type heuristic argument as to why this particular PDE should model heat flow. In a region in $\mathbf{R}^3$ with the usual coordinates x, y, z, let

$$u(x, y, z, t) = \text{temperature at time } t \text{ at } (x, y, z).$$

Definition 14.4.1 *The* heat equation *is:*

$$\frac{\partial^2 u}{\partial x^2} + \frac{\partial^2 u}{\partial y^2} + \frac{\partial^2 u}{\partial z^2} = c\frac{\partial u}{\partial t},$$

where c is a constant.

Frequently one starts with an initial specified temperature distribution, such as

$$u(x, y, z, 0) = f(x, y, z),$$

with $f(x, y, z)$ some known, given function.

Surprisingly, the heat equation shows up throughout mathematics and the sciences, in many contexts for which no notion of heat or temperature is apparent. The common theme is that heat is a type of diffusion process and that the heat equation is the PDE that will capture any diffusion process. Also, there are a number of techniques for solving the heat equation. In fact, using Fourier Analysis, we solved it in the one-dimensional case in Chapter Thirteen. The method of separation of variables, used in last section to solve the Laplacian, can also be used.

Now to see why the above PDE deserves the name 'heat equation'. As seen in the last section,

$$\triangle u = \frac{\partial^2 u}{\partial x^2} + \frac{\partial^2 u}{\partial y^2} + \frac{\partial^2 u}{\partial z^2}$$

is the Laplacian. In non-rectilinear coordinates, the Laplacian will have different looking forms, but the heat equation will always be:

$$\triangle u = c\frac{\partial u}{\partial t}.$$

For simplicity, we restrict ourselves to the one-dimensional case. Consider an infinitely long rod, which we denote by the x-axis.

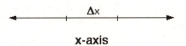

x-axis

Though the basic definitions of heat and temperature are and were fraught with difficulties, we will assume that there is a notion of temperature and that heat is measured via the change in temperature. Let $u(x, t)$ denote the temperature at position x at time t. We now denote the change in a variable by $\triangle u$, $\triangle x$, $\triangle t$, etc. Note that here $\triangle$ is not denoting the Laplacian of these variables.

There are three important constants associated to our rod, all coming from the real world: the density ρ, the thermal conductivity k and the specific heat σ. The density arises in that the mass m of the rod over a distance $\triangle x$ will be the product $\rho \cdot \triangle x$. The specific heat is the number σ that, if a length $\triangle x$ of the rod has its temperature u raised to $u + \triangle u$, then its heat will change by $\sigma \cdot (mass) \cdot \Delta u$. Note that this last number is the same as $\sigma \cdot \rho \cdot \Delta x \cdot \Delta u$. Here we are using the notion that heat is a measure of the change in temperature. Finally, the thermal conductivity k is the constant that yields

$$k \cdot \frac{\Delta u}{\Delta x}\Big|_x$$

as the amount of heat that can flow through the rod at a fixed point x. Via physical experiments, these constants can be shown to exist.

We want to see how much heat flows in and out of the interval $[x, x+\Delta x]$. By calculating this heat flow by two different methods, and then letting $\Delta x \to 0$, the heat equation will appear. First, if the temperature changes by Δu, the heat will change by

$$\sigma \cdot \rho \cdot \Delta x \cdot \Delta u.$$

Second, at the point $x + \Delta x$, the amount of heat flowing out will be, over time Δt,

$$k \cdot \frac{\Delta u}{\Delta x}\Big|_{x+\Delta x}\Delta t.$$

$-k \dfrac{\Delta u}{\Delta x}\Big|_x$ Δt = heat flow out x end

$$\cdots - - - + \underset{\mathbf{x}}{\underline{\qquad \Delta \mathbf{x} \qquad}} + \cdots \cdots$$
$$\mathbf{x} \qquad\qquad \mathbf{x} + \Delta \mathbf{x}$$

$k \dfrac{\Delta u}{\Delta x}\Big|_{x+\Delta x}$ Δt = heat flow out x+Δx end

At the point x, the amount of heat flowing out will be, over time Δt,

$$-k \cdot \frac{\Delta u}{\Delta x}\Big|_x\Delta t.$$

Then the heat change over the interval Δx will also be

$$(k\frac{\Delta u}{\Delta x}\Big|_{x+\Delta x} - k\frac{\Delta u}{\Delta x}\Big|_x)\Delta t.$$

Thus

$$k \cdot (\frac{\Delta u}{\Delta x}\Big|_{x+\Delta x} - \frac{\Delta u}{\Delta x}\Big|_x)\Delta t = \sigma\rho\Delta x\Delta u.$$

Then

$$\frac{(\frac{\Delta u}{\Delta x}|_{x+\Delta x} - \frac{\Delta u}{\Delta x}|_{x})}{\Delta x} = \frac{\sigma\rho}{k}\frac{\Delta u}{\Delta t}.$$

Letting Δx and Δt approach 0, we get by the definition of partial differentiation the heat equation

$$\frac{\partial^2 u}{\partial x^2} = \frac{\sigma\rho}{k} \cdot \frac{\partial u}{\partial t}.$$

In fact, we see that the constant c is

$$c = \frac{\sigma\rho}{k}.$$

Again, there are at least two other methods for solving the heat equation. We can, for example, use Fourier transforms, which is what we used to solve it in Chapter Thirteen. We can also use the method of separation of variables, discussed in the previous section.

14.5 The Wave Equation

14.5.1 Derivation

As its name suggests, this partial differential equation was originally derived to describe the motion of waves. As with the heat equation, its basic form appears in many apparently non-wave-like areas. We will state the wave equation and then give a quick heuristic description of why the wave equation should describe waves.

A transverse wave in the $x - y$ plane travelling in the x-direction should look like:

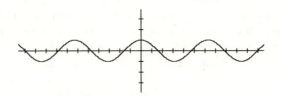

The solution function is denoted by $y(x, t)$, which is just the y coordinate of the wave at place x at time t. The wave equation in two independent variables is

$$\frac{\partial^2 y}{\partial x^2} - c\frac{\partial^2 y}{\partial t^2} = 0,$$

where c is a positive number. Usually we start with some type of knowledge of the initial position of the wave. This will of course mean that we are given an initial function $f(x)$ such that

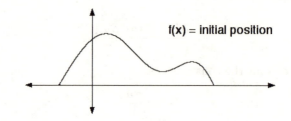

f(x) = initial position

$$y(x, 0) = f(x).$$

In general, the wave equation in n variables $x_1, \ldots, x_n$ with initial condition $f(x_1, \ldots, x_n)$ is

$$\frac{\partial^2 y}{\partial x_1{}^2} + \ldots + \frac{\partial^2 y}{\partial x_n{}^2} - c \cdot \frac{\partial^2 y}{\partial t^2} = 0$$

with initial condition

$$y(x_1, \ldots, x_n, 0) = f(x_1, \ldots, x_n).$$

In nonrectilinear coordinates, the wave equation will be:

$$\triangle y(x_1, \ldots, x_n, t) - c \cdot \frac{\partial^2 y}{\partial t^2} = 0.$$

Now to see the heuristics behind why this partial differential equation is even called the wave equation. Of course we need to make some physical assumptions. Assume that the wave is a string moving in an 'elastic' medium, meaning that subject to any displacement, there is a restoring force, something trying to move the string back to where it was. We further assume that the initial disturbance is small. We will use that

$$\text{Force} = (\text{mass}) \cdot (\text{acceleration}).$$

We let our string have density ρ and assume that there is a tension T in the string (this tension will be what we call the restoring force) which will act tangentially on the string. Finally, we assume that the string can only move vertically.

Consider the wave

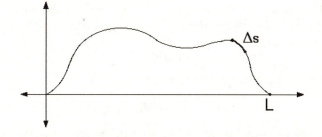

Let s denote the arc length of the curve. We want to calculate the restoring force acting on the segment $\triangle s$ of the curve in two different ways and then let $\triangle s \to 0$. Since the density is ρ, the mass of the segment $\triangle s$ will be the product $(\rho \cdot \triangle s)$. The acceleration is the second derivative. Since we are assuming that the curve can only move vertically (in the y-direction), the acceleration will be $\frac{\partial^2 y}{\partial t^2}$. Thus the force will be

$$(\rho \cdot \triangle s) \cdot \frac{\partial^2 y}{\partial t^2}.$$

By the assumption that the displacement is small, we can approximate the arc length $\triangle s$ by the change in the x-direction alone.

$$\triangle s \sim \triangle x$$

Hence we assume that the restoring force is

$$(\rho \triangle x) \cdot \frac{\partial^2 y}{\partial t^2}.$$

Now to calculate the restoring force in a completely different way. At each point in the picture

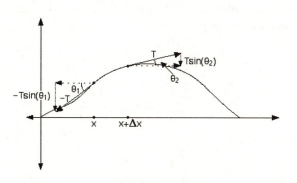

the tension T gives rise to an acceleration tangent to the curve. We want the y component. At the point $x + \triangle x$, the restoring force will be

$$T \sin\theta_2.$$

At the point x, the restoring force will be

$$-T \sin\theta_1.$$

Since both angles θ_1 and θ_2 are small, we can use the following approximation

$$\sin\theta_1 \quad \sim \quad \tan\theta_1 = \frac{\partial y}{\partial x}|_x.$$

$$\sin\theta_2 \quad \sim \quad \tan\theta_2 = \frac{\partial y}{\partial x}|_{x+\triangle x}.$$

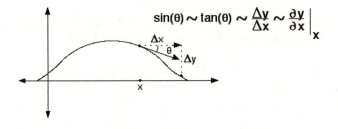

Then we can set the restoring force to be

$$T(\frac{\partial y}{\partial x}|_{x+\triangle x} - \frac{\partial y}{\partial x}|_x).$$

As we have now calculated the restoring force in two different ways, we can set the two formulas equal:

$$T(\frac{\partial y}{\partial x}|_{x+\triangle x} - \frac{\partial y}{\partial x}|_x) = \rho\triangle x \cdot \frac{\partial^2 y}{\partial t^2}.$$

or

$$\frac{\frac{\partial y}{\partial x}|_{x+\triangle x} - \frac{\partial y}{\partial x}|_x}{\triangle x} = \frac{\rho}{T}\frac{\partial^2 y}{\partial t^2}.$$

Letting $\triangle x \to 0$, we get

$$\frac{\partial^2 y}{\partial x^2} = \frac{\rho}{T}\frac{\partial^2 y}{\partial t^2},$$

the wave equation.

Now to see what solutions look like. We assume that $y(0) = 0$ and $y(L) = 0$, for some constant L. Thus we restrict our attention to waves which have fixed endpoints.

An exercise at the end of the chapter will ask you to solve the wave equation using the method of separation of variables and via Fourier transforms. Your answer will in fact be:

$$y(x,t) = \sum_{n=1}^{\infty} k_n sin\left(\frac{n\pi x}{L}\right) cos\left(\frac{n\pi t}{L}\right)$$

where

$$k_n = \frac{2}{L} \int_0^L f(x) sin\left(\frac{n\pi x}{L}\right) dx.$$

14.5.2 Change of Variables

Sometimes a clever change of variables will reduce the original PDE to a more manageable one. We will see this in the following solution of the wave equation. Take an infinitely long piece of string. Suppose we pluck the string in the middle and then let go.

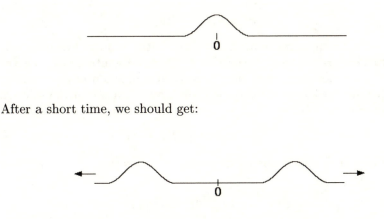

After a short time, we should get:

with seemingly two waves moving in opposite directions but at the same speed. With much thought and cleverness, one might eventually try to change coordinate systems in an attempt to capture these two waves.

Thus suppose we want to solve

$$\frac{\partial^2 y}{\partial x^2} - \frac{1}{c^2}\frac{\partial^2 y}{\partial t^2} = 0,$$

subject to the initial conditions

$$y(x,0) = g(x) \text{ and } \frac{\partial y}{\partial t}(x,0) = h(x)$$

for given functions $g(x)$ and $h(x)$. Note that we have relabelled the constant in the wave equation to be $\frac{1}{c^2}$. This is done solely for notational convenience, as we will in a moment.

Now to make the change of variables. Set

$$u = x + ct \text{ and } v = x - ct.$$

Using the chain rule, this coordinate change transforms the original wave equation into:

$$\frac{\partial^2 y}{\partial u \partial v} = 0.$$

We can solve this PDE by two straightforward integrations. First integrate with respect to the variable u to get

$$\frac{\partial y}{\partial v} = a(v),$$

where $a(v)$ is an unknown function of the variable v alone. This new function $a(v)$ is the 'constant of integration', constant with respect to the u variable. Now integrate this with respect to v to get

$$y(u,v) = A(v) + B(u),$$

where $A(v)$ is the integral of $a(v)$ and $B(u)$ is the term representing the 'constant of integration' with respect to v. Thus the solution $y(u,v)$ is the sum of two, for now unknown, functions, each a function of one variable alone. Plugging back into our original coordinates means that the solution will have the form:

$$y(u,v) = A(x - ct) + B(x + ct).$$

We use our initial conditions to determine the functions $A(x - ct)$ and $B(x + ct)$. We have

$$g(x) = y(x,0) = A(x) + B(x)$$

and

$$h(x) = \frac{\partial y}{\partial t}(x,0) = -cA'(x) + cB'(x).$$

For this last equation, integrate with respect to the one variable x, to get that

$$\int_0^x h(s)\mathrm{d}s + C = -cA(x) + cB(x).$$

Since we are assuming that the functions $g(x)$ and $h(x)$ are known, we can now solve for $A(x)$ and $B(x)$, to get:

$$A(x) = \frac{1}{2}g(x) - \frac{1}{2c}\int_0^x h(s)\mathrm{d}s - \frac{C}{2c}$$

and

$$B(x) = \frac{1}{2}g(x) + \frac{1}{2c}\int_0^x h(s)\mathrm{d}s + \frac{C}{2c}.$$

Then the solution is:

$$
\begin{aligned}
y(x,t) &= A(x-ct) + B(x+ct) \\
&= \frac{g(x-ct) + g(x+ct)}{2} + \frac{1}{2c}\int_{x-ct}^{x+ct} h(s)\mathrm{d}s,
\end{aligned}
$$

This is called the *d'Alembert formula*. Note that if the initial velocity $h(x) = 0$, then the solution is simply

$$y(x,t) = \frac{g(x-ct) + g(x+ct)}{2},$$

which is two waves travelling in opposite directions, each looking like the initial position. (Though this is a standard way to solve the wave equation, I took the basic approach from Davis' *Fourier Series and Orthogonal Functions* [24].)

This method leaves the question of how to find a good change of coordinates unanswered. This is an art, not a science.

14.6 The Failure of Solutions: Integrability Conditions

There are no known general methods for determining when a system of partial differential equations has a solution. Frequently, though, there are necessary conditions (usually called 'integrability conditions') for there to be a solution.

We will look at the easiest case. When will there be a two-variable function $f(x,y)$, defined on the plane $\mathbf{R}^2$, satisfying:

$$\frac{\partial f}{\partial x} = g_1(x,y)$$

and

$$\frac{\partial f}{\partial y} = g_2(x,y),$$

where both g_1 and g_2 are differentiable functions? In this standard result from multivariable calculus, there are clean necessary and sufficient conditions for the solution function f to exist:

Theorem 14.6.1 *There is a solution f to the above system of partial differential equations if and only if*

$$\frac{\partial g_1}{\partial y} = \frac{\partial g_2}{\partial x}.$$

In this case, the *integrability* condition is $\frac{\partial g_1}{\partial y} = \frac{\partial g_2}{\partial x}$. As we will see, this is the easy part of the theorem; it is also the model for integrability conditions in general.

Proof: First assume that we have our solution f satisfying $\frac{\partial f}{\partial x} = g_1(x, y)$ and $\frac{\partial f}{\partial y} = g_2(x, y)$. Then

$$\frac{\partial g_1}{\partial y} = \frac{\partial}{\partial y}\frac{\partial f}{\partial x} = \frac{\partial}{\partial x}\frac{\partial f}{\partial y} = \frac{\partial g_2}{\partial x}.$$

Thus the integrability condition is just a consequence that the order for taking partial derivatives does not matter.

The other direction takes more work. As a word of warning, Green's Theorem will be critical. We must find a function $f(x, y)$ satisfying the given system of PDEs. Given any point (x, y) in the plane, let γ be any smooth path from the origin $(0, 0)$ to (x, y). Define

$$f(x, y) = \int_\gamma g_1(x, y)\mathrm{d}x + g_2(x, y)\mathrm{d}y.$$

We first show that the function $f(x, y)$ is well-defined, meaning that its value is independent of which path γ is chosen. This will then allow us to show that $\frac{\partial f}{\partial x} = g_1(x, y)$ and $\frac{\partial f}{\partial y} = g_2(x, y)$. Let τ be another smooth path from $(0, 0)$ to (x, y).

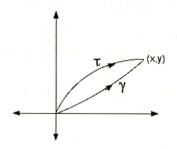

We want to show that

$$\int_\gamma g_1(x,y)\mathrm{d}x + g_2(x,y)\mathrm{d}y = \int_\tau g_1(x,y)\mathrm{d}x + g_2(x,y)\mathrm{d}y.$$

We can consider $\gamma - \tau$ as a closed loop at the origin, enclosing a region R. (Note: it might be the case that $\gamma - \tau$ encloses several regions, but then just apply the following to each of these regions.) By Green's Theorem we have

$$
\begin{aligned}
\int_\gamma g_1\mathrm{d}x + g_2\mathrm{d}y - \int_\tau g_1\mathrm{d}x + g_2\mathrm{d}y &= \int_{\gamma-\tau} g_1\mathrm{d}x + g_2\mathrm{d}y \\
&= \int_R (\frac{\partial g_2}{\partial x} - \frac{\partial g_1}{\partial y})\mathrm{d}x\mathrm{d}y \\
&= 0
\end{aligned}
$$

by the assumption that $\frac{\partial g_1}{\partial y} = \frac{\partial g_2}{\partial x}$. Thus the function $f(x,y)$ is well-defined.

Now to show that this function f satisfies $\frac{\partial f}{\partial x} = g_1(x,y)$ and $\frac{\partial f}{\partial y} = g_2(x,y)$. We will just show the first, as the second is similar. The key is that we will reduce the problem to the Fundamental Theorem of Calculus. Fix a point (x_0, y_0). Consider any path γ from $(0,0)$ to (x_0, y_0) and the extension $\gamma' = \gamma + \tau$, where τ is the horizontal line from (x_0, y_0) to (x, y_0).

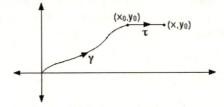

Then

$$
\begin{aligned}
\frac{\partial f}{\partial x} &= \lim_{x \to x_0} \frac{f(x, y_0) - f(x_0, y_0)}{x - x_0} \\
&= \lim_{x \to x_0} \frac{\int_{x_0}^x g_1(t, y_0)\mathrm{d}t}{x - x_0},
\end{aligned}
$$

since there is no variation in the y-direction, forcing the g_2 part of the path integral to drop out. This last limit, by the Fundamental Theorem of Calculus, is equal to g_1, as desired. $\square$

14.7 Lewy's Example

Once you place any natural integrability conditions on a system of partial differential equations, you can then ask if there will always be a solution.

In practice, often such general statements about the existence of solutions can be made. For example, in the middle of the twentieth century it was shown that given any complex numbers $a_1, \ldots, a_n$ and any smooth function $g(x_1, \ldots, x_n)$, there always exists a smooth solution $f(x_1, \ldots, x_n)$ satisfying

$$a_1 \frac{\partial f}{\partial x_1} + \ldots + a_n \frac{\partial f}{\partial x_n} = g.$$

Based in part on these types of results, it was the belief that all reasonable PDEs would have solutions. Then, in 1957, Hans Lewy showed the amazing result that the linear PDE

$$\frac{\partial f}{\partial x} + i\frac{\partial f}{\partial y} - (x + iy)\frac{\partial f}{\partial z} = g(x, y, z)$$

will have a solution f only if g is real-analytic. Note that while this PDE does not have constants as coefficients, the coefficients are about as reasonable as you could want. Lewy's proof, while not hard (see Folland's book on PDEs [39]), did not give any real indication as to why there is no solution. In the early 1970s, Nirenberg showed that the Lewy PDE did not have a solution due to that there existed a three-dimensional CR structure (a certain type of manifold) that could not be embedded into a complex space, thus linking a geometric condition to the question of existence of this PDE. This is a common tack, namely to concentrate on PDEs whose solutions have some type of geometric meaning. Then, in trying to find the solution, use the geometry as a guide.

14.8 Books

Since beginning differential equations is a standard sophomore level course, there are many beginning text books. Boyce and Diprima's book [12] has long been a standard. Simmon's book [99] is also good. Another approach to learning basic ODEs is to volunteer to TA or teach such a class (though I would recommend that you teach linear algebra and vector calculus first). Moving into the realm of PDEs the level of text becomes much harder and more abstract. I have learned a lot from Folland's book [39]. Fritz John's book [69] has long been a standard. I have heard that Evans' recent book [33] is also excellent.

14.9 Exercises

1. The most basic differential equation is probably

$$\frac{dy}{dx} = y,$$

subject to the boundary condition $y(0) = 1$. The solution is of course the exponential function $y(x) = e^x$. Use Picard iteration to show that this is indeed the solution to $\frac{dy}{dx} = y$. (Of course you get an answer as a power series and then need to recognize that the power series is e^x. The author realizes that if you know the power series for the exponential function you also know that it is its own derivative. The goal of this problem is see explicitly how Picard iteration works on the simplest possible differential equation.)

2. Let $f(x)$ be a one variable function, with domain the interval $[0, 1]$, whose first derivative is continuous. Show that f is Lipschitz.

3. Show that $f(x) = e^x$ is not Lipschitz on the real numbers.

4. Solve the wave equation

$$\frac{\partial^2 y}{\partial x^2} - c\frac{\partial^2 y}{\partial t^2} = 0$$

subject to the boundary conditions $y(0, t) = 0$ and $y(L, t) = 0$ and the initial condition $y(x, 0) = f(x)$ for some function $f(x)$.

a. Use the method of separation of variables as described in the section on the Laplacian.

b. Now find the solutions using Fourier transforms.

Chapter 15

Combinatorics and Probability Theory

Basic Goals:	Cleverly Counting Large Finite Sets
	Central Limit Theorem

Beginning probability theory is basically the study of how to count large finite sets, or in other words, an application of combinatorics. Thus the first section of this chapter deals with basic combinatorics. The next three sections deal with the basics of probability theory. Unfortunately, counting will only take us so far in probability. If we want to see what happens as we, for example, play a game over and over again, methods of calculus become important. We concentrate on the Central Limit Theorem, which is where the famed Gauss-Bell curve appears. The proof of the Central Limit Theorem is full of clever estimates and algebraic tricks. We include this proof not only due to the importance of the Central Limit Theorem but also to show people that these types of estimates and tricks are sometimes needed in mathematics.

15.1 Counting

There are many ways to count. The most naive method, the one we learn as children, is simply to explicitly count the elements in a set, and this method is indeed the best one for small sets. Unfortunately, many sets are just too large for anyone to merely count the elements. Certainly in large part the fascination in card games such as poker and bridge is that while there are only a finite number of possible hands, the actual number is far too large for anyone to deal with directly, forcing the players to develop

strategies and various heuristical devices. Combinatorics is the study of
how to cleverly count. Be warned that the subject can quickly get quite
difficult and is becoming increasingly important in mathematics.

We will look at the simplest of combinatorial formulas, ones that have
been known for centuries. Start with n balls. Label each ball with a
number $1, 2, \ldots, n$ and then put the balls into an urn. Pull one out, record
its number and then put the ball back in. Again, pull out a ball and record
its number and put it back into the urn. Keep this up until k balls have
been pulled out and put back into the urn. We want to know how many
different k-tuples of numbers are possible.

To pull out two balls from a three-ball urn (here $n = 3$ and $k = 2$), we
can just list the possibilities:

$$(1,1), (1,2), (1,3), (2,1), (2,2), (2,3), (3,1), (3,2), (3,3).$$

But if we pull out seventy-six balls from a ninety-nine ball urn (here $n = 99$
and $k = 76$), it would be ridiculous to make this list.

Nevertheless, we can find the correct number. There are n possibilities
for the first number, n possibilities for the second, n for the third, etc. Thus
all told there must be n^k possible ways to choose k-tuples of n numbers.
This is a formula that works no matter how many balls we have or how
many times we choose a ball.

For the next counting problem, return to the urn. Pull out a ball, record
its number and keep it out. Now pull out another ball, record its number
and keep it out. Continue pulling out balls and not replacing them. Now
we want to find out how many k-tuples of n numbers there are without
replacement. There are n possibilities for the first number, only $(n - 1)$
possibilities for the second, $(n - 2)$ for the third, etc. Thus the number of
ways of choosing from n balls k times without replacement is:

$$n(n - 1)(n - 2) \cdots (n - k + 1).$$

For our next counting problem, we want to find out how many ways
there are for pulling out k balls from an urn with n balls, but now not only
not replacing the balls but also not caring about the order of the balls.
Thus pulling out the balls $(1, 2, 3)$ will be viewed as equivalent to pulling
out the balls $(2, 1, 3)$. Suppose we have already pulled out k of the balls.
We want to see how many ways there are of mixing up these k balls. But
this should be the same as how many ways are there of choosing from k
balls k times, which is

$$k(k - 1)(k - 2) \cdots 2 \cdot 1 = k!.$$

Since $n(n-1)(n-2) \cdots (n-k+1)$ is the number of ways of choosing from n
balls k times with order mattering and with each ordering capable of being

mixed up $k!$ ways, we have

$$\frac{n(n-1)\ldots(n-k+1)}{k!} = \frac{n!}{k!(n-k)!},$$

which is the number of ways of choosing k balls from n balls without replacement and with order not mattering. This number comes up so often it has its own symbol

$$\binom{n}{k} = \frac{n!}{k!(n-k)!},$$

pronounced 'n choose k'. It is frequently called the *binomial coefficient*, due to its appearance in the Binomial Theorem:

$$(a+b)^n = \sum_{k=0}^{n} \binom{n}{k} a^k b^{n-k}.$$

The idea is that $(a+b)^n = (a+b)(a+b)\ldots(a+b)$. To calculate how many different terms of the form $a^k b^{n-k}$ we can get, we note that this is the same as counting how many ways we can choose k things from n things without replacement and with ordering not mattering.

15.2 Basic Probability Theory

We want to set up the basic definitions of elementary probability theory. These definitions are required to yield the results we all know, such as that there is a fifty-fifty chance of flipping a coin and getting heads, or that there is a one in four chance of drawing a heart from a standard deck of 52 cards. Of course, as always, the reason for worrying about the basic definitions is not just to understand the obvious odds of getting heads but that the correct basic definition will allow us to compute the probabilities of events that are quite complicated.

We start with the notion of a sample space ω, which technically is just another name for a set. Intuitively, a sample space ω is the set whose elements are what can happen, or more precisely, the possible outcomes of an event. For example, if we flip a coin twice, ω will be a set with the four elements

$$\{(heads, heads), (heads, tails), (tails, heads), (tails, tails)\}.$$

Definition 15.2.1 *Let ω be a sample space and A a subset of ω. Then the probability of A, denoted by $P(A)$, is the number of elements in A divided by the number of elements in the sample space ω. Thus*

$$P(A) = \frac{|A|}{|\omega|},$$

where $|A|$ denotes the number of elements in the set A.

For example, if

$$\omega = \{(heads, heads), (heads, tails), (tails, heads), (tails, tails)\},$$

and if $A = \{(heads, heads)\}$, then the probability of flipping a coin twice and getting two heads will be

$$P(A) = \frac{|A|}{|\omega|} = \frac{1}{4},$$

which agrees with common sense.

In this framework, many of the basic rules of probability reduce to rules of set theory. For example, via sets, we see that

$$P(A \cup B) = P(A) + P(B) - P(A \cap B).$$

Frequently, a subset A of a sample space ω is called an *event*.

There are times when it is too much trouble to actually translate a real-world probability problem into a question of size of sets. For example, suppose we are flipping an unfair coin, where there is a 3/4 chance of getting a head and a 1/4 chance of getting tails. We could model this by taking our sample set to be

$$\omega = \{heads_1, heads_2, heads_3, tails\},$$

where we are using subscripts to keep track of the different ways of getting heads, but this feels unnatural. A more natural sample space would be

$$\omega = \{heads, tails\},$$

and to somehow account for the fact that it is far more likely to get heads than tails. This leads to another definition of a probability space:

Definition 15.2.2 *A* probability space *is a set ω, called the* sample space, *and a function*

$$P : \omega \to [0, 1]$$

such that

$$\sum_{a \in \omega} P(a) = 1.$$

We say that the probability of getting an 'a' is the value of $P(a)$.

If on a sample space ω it is equally likely to get any single element of ω, i.e., for all $a \in \omega$ we have

$$P(a) = \frac{1}{|\omega|},$$

then our 'size of set' definition for probability will agree with this second definition. For the model of flipping an unfair coin, this definition will give us that the sample set is:

$$\omega = \{heads, tails\},$$

but that $P(heads) = 3/4$ and $P(tails) = 1/4$.

We now turn to the notion of a random variable.

Definition 15.2.3 *A* random variable **X** *on a sample space ω is a real-valued function on ω:*

$$\mathbf{X} : \omega \to \mathbf{R}.$$

For example, we now create a simplistic gambling game which requires two flips of a coin. Once again let the sample space be

$$\omega = \{(heads, heads), (heads, tails), (tails, heads), (tails, tails)\}.$$

Suppose that, if the first toss of a coin is heads, you win ten dollars. If it is tails, you lose five dollars. On the second toss, heads will pay fifteen dollars and tails will cost you twelve dollars. To capture these stakes (for an admittedly boring game), we define the random variable

$$\mathbf{X} : \omega \to \mathbf{R}$$

by

$$\mathbf{X}(heads, heads) = 10 + 15 = 25$$

$$\mathbf{X}(heads, tails) = 10 - 12 = -2$$

$$\mathbf{X}(tails, heads) = -5 + 15 = 10$$

$$\mathbf{X}(tails, tails) = -5 - 12 = -17.$$

15.3   Independence

Toss a pair of dice, one blue and one red. The number on the blue die should have nothing to do with the the number on the red die. The events are in some sense independent, or disjoint. We want to take this intuition of independence and give it a sharp definition.

Before giving a definition for independence, we need to talk about conditional probability. Start with a sample space ω. We want to understand the probability for an event A to occur, given that we already know some other event B has occurred. For example, roll a single die. Let ω be the six possible outcomes on this die. Let A be the event that a 4 shows up. Certainly we have

$$P(A) = \frac{|A|}{|\omega|} = \frac{1}{6}.$$

But suppose someone tells us, before we look at the rolled die, that they know for sure that on the die there is an even number. Then the probability that a 4 will occur should be quite different. The set $B = \{2, 4, 6\}$ is the event that an even number occurs. Then the probability that a 4 shows up should now be $1/3$, as there are only three elements in B. Note that

$$\frac{1}{3} = \frac{|A \cap B|}{|B|} = \frac{\frac{|A \cap B|}{|\omega|}}{\frac{|B|}{|\omega|}} = \frac{P(A \cap B)}{P(B)}.$$

This motivates the definition:

Definition 15.3.1 *The* conditional probability *that A occurs given that B has occurred is:*

$$P(A|B) = \frac{P(A \cap B)}{P(B)}.$$

What should it mean for an event A to be independent from an event B? At the least, it should mean that knowing about the likelihood of event B occurring should have no bearing on the likelihood that A occurs, i.e., knowing about B should not effect A. Thus if A and B are independent, we should have

$$P(A|B) = P(A).$$

Using that $P(A|B) = \frac{P(A \cap B)}{P(B)}$, this means that a reasonable definition for independence is:

Definition 15.3.2 *Two events A and B are* independent *if*

$$P(A \cap B) = P(A) \cdot P(B).$$

15.4 Expected Values and Variance

In a game, how much should you be expected to win in the long run? This quantity is the expected value. Further, how likely is it that you might lose big time, even if the expected value tells you that you will usually come out ahead? This type of information is contained in the variance and in its square root, the standard deviation. We start with some definitions.

Definition 15.4.1 *The* expected value *of a random variable* $\mathbf{X}$ *on a sample space* ω *is:*

$$E(\mathbf{X}) = \sum_{a \in \omega} \mathbf{X}(a) \cdot P(a).$$

For example, recall the simplistic game defined at the end of section two, where we flip a coin twice and our random variable represents our winnings: $\mathbf{X}(heads, heads) = 10 + 15 = 25, \mathbf{X}(heads, tails) = 10 - 12 = -2, \mathbf{X}(tails, heads) = -5 + 15 = 10$, and $\mathbf{X}(tails, tails) = -5 - 12 = -17$. The expected value is simply:

$$
\begin{aligned}
E(\mathbf{X}) &= 25 \left(\frac{1}{4}\right) + (-2)\left(\frac{1}{4}\right) + 10 \left(\frac{1}{4}\right) + (-17)\left(\frac{1}{4}\right) \\
&= 4.
\end{aligned}
$$

Intuitively, this means that on average you will win four dollars each time you play the game. Of course, luck might be against you and you could lose quite a bit.

The expected value can be viewed as a function from the set of all random variables to the real numbers. As a function, the expected value is linear.

Theorem 15.4.1 *On a probability space, the expected value is linear, meaning that for all random variables* $\mathbf{X}$ *and* $\mathbf{Y}$ *and all real numbers* λ *and* μ, *we have*

$$E(\lambda \mathbf{X} + \mu \mathbf{Y}) = \lambda E(\mathbf{X}) + \mu E(\mathbf{Y}).$$

Proof: This is a straightforward calculation from the definition of expected value. We have

$$
\begin{aligned}
E(\lambda \mathbf{X} + \mu \mathbf{Y}) &= \sum_{a \in \omega}(\lambda \mathbf{X} + \mu \mathbf{Y})(a) \cdot P(a) \\
&= \sum_{a \in \omega}(\lambda \mathbf{X}(a) + \mu \mathbf{Y}(a)) \cdot P(a) \\
&= \sum_{a \in \omega} \lambda \mathbf{X}(a) \cdot P(a) + \sum_{a \in \omega} \mu \mathbf{Y}(a) \cdot P(a)
\end{aligned}
$$

$$= \lambda \sum_{a \in \omega} \mathbf{X}(a) \cdot P(a) + \mu \sum_{a \in \omega} \mathbf{Y}(a) \cdot P(a)$$

$$= \lambda E(\mathbf{X}) + \mu E(\mathbf{Y}). \ \square$$

The expected value will only tell a part of the story, though. Consider two classes, each with ten students. On a test, in one of the classes five people got 100s and five got 50s, while in the other everyone got a 75. In both classes the average was a 75 but the performances were quite different. Expected value is like the average, but it does not tell us how far from the average you are likely to be. For example, in the first class you are guaranteed to be 25 points from the average while in the second class you are guaranteed to be exactly at the average. There is a measure of how likely it is that you are far from the expected value:

Definition 15.4.2 *The* variance *of a random variable* $\mathbf{X}$ *on a sample space* ω *is*

$$V(\mathbf{X}) = E[\mathbf{X} - E(\mathbf{X})]^2.$$

The idea is we set up a new random variable,

$$[\mathbf{X} - E(\mathbf{X})]^2.$$

Note that the expected value $E(\mathbf{X})$ is just a number. The farther $\mathbf{X}$ is from its expected value $E(\mathbf{X})$, the larger is $[\mathbf{X} - E(\mathbf{X})]^2$. Thus it is a measure of how far we can be expected to be from the average. We square $\mathbf{X} - E(\mathbf{X})$ in order to make everything non-negative.

We can think of the variance V as a map from random variables to the real numbers. While not quite linear, it is close, as we will now see. First, though, we want to show that the formula for variance can be rewritten.

Lemma 15.4.1 *For a random variable* $\mathbf{X}$ *on a probability space, we have*

$$V(\mathbf{X}) = E(\mathbf{X}^2) - [E(\mathbf{X})]^2$$

Proof: This is a direct calculation. We are interested in the new random variable

$$[\mathbf{X} - E(\mathbf{X})]^2.$$

Now

$$[\mathbf{X} - E(\mathbf{X})]^2 = \mathbf{X}^2 - 2\mathbf{X}E(\mathbf{X}) + [E(\mathbf{X})]^2.$$

Since $E(\mathbf{X})$ is just a number and since the expected value, as a map from random variables to the reals, is linear, we have

$$\begin{aligned} V(\mathbf{X}) &= E[\mathbf{X} - E(\mathbf{X})]^2 \\ &= E[\mathbf{X}^2 - 2\mathbf{X}E(\mathbf{X}) + [E(\mathbf{X})]^2] \\ &= E(\mathbf{X}^2) - 2E(\mathbf{X})E(\mathbf{X}) + [E(\mathbf{X})]^2 \\ &= E(\mathbf{X}^2) - [E(\mathbf{X})]^2, \end{aligned}$$

as desired. $\Box$

This will allow us to show that the variance is almost linear.

Theorem 15.4.2 *Let* $\mathbf{X}$ *and* $\mathbf{Y}$ *be any two random variables that are independent on a probability space and let* λ *be any real number. Then*

$$V(\lambda \mathbf{X}) = \lambda^2 V(\mathbf{X})$$

and

$$V(\mathbf{X} + \mathbf{Y}) = V(\mathbf{X}) + V(\mathbf{Y}).$$

It is the λ^2 term that prevents the variance from being linear.

Proof: Since the expected value is a linear function, we know that $E(\lambda \mathbf{X}) = \lambda E(\mathbf{X})$. Then

$$
\begin{aligned}
V(\lambda \mathbf{X}) &= E[(\lambda \mathbf{X})^2] - [E(\lambda \mathbf{X})]^2 \\
&= \lambda^2 E(\mathbf{X}^2) - [\lambda E(\mathbf{X})]^2 \\
&= \lambda^2 [E(\mathbf{X}^2) - [E(\mathbf{X})]^2] \\
&= \lambda^2 V(\mathbf{X}).
\end{aligned}
$$

For the second formula, we will need to use that the independence of $\mathbf{X}$ and $\mathbf{Y}$ means that

$$E(\mathbf{XY}) = E(\mathbf{X})E(\mathbf{Y}).$$

By the above lemma's description of variance, we have

$$
\begin{aligned}
V(\mathbf{X} + \mathbf{Y}) &= E[(\mathbf{X} + \mathbf{Y})^2] - [E(\mathbf{X} + \mathbf{Y})]^2 \\
&= E[\mathbf{X}^2 + 2\mathbf{XY} + \mathbf{Y}^2] - [E(\mathbf{X}) + E(\mathbf{Y})]^2 \\
&= E[\mathbf{X}^2] + 2E[\mathbf{XY}] + E[\mathbf{Y}^2] \\
&\quad -[E(\mathbf{X})]^2 - 2E(\mathbf{X})E(\mathbf{Y}) - [E(\mathbf{Y})]^2 \\
&= (E[\mathbf{X}^2] - [E(\mathbf{X})]^2) + (2E[\mathbf{XY}] \\
&\quad -2E(\mathbf{X})E(\mathbf{Y})) + (E[\mathbf{Y}^2] - [E(\mathbf{Y})]^2) \\
&= V(\mathbf{X}) + V(\mathbf{Y}),
\end{aligned}
$$

as desired. $\Box$

A number related to the variance is its square root, the *standard deviation*:

$$\text{standard deviation}(\mathbf{X}) = \sigma(\mathbf{X}) = \sqrt{V(\mathbf{X})}.$$

15.5   Central Limit Theorem

In the last section we defined the basic notions of probability in terms of counting. Unfortunately, combinatorics can only take us so far. Think about flipping a coin. After many flips, we expect that the total number of heads should be quite close to one half of the total number of flips. In trying to capture this notion of flipping a coin over and over again, we need to introduce the following:

Definition 15.5.1 *Repeated independent trials are called* Bernoulli trials *if there are only two possible outcomes for each trial and if their probabilities remain the same throughout the trials.*

Let A be one of the outcomes and suppose the probability of A is $P(A) = p$. Then the probability of A not occurring is $1 - p$, which we will denote by q. Let the sample space be

$$\omega = \{A, \text{not } A\}.$$

We have

$$P(A) = p, P(\text{not } A) = q.$$

We now want to see what happens when we take many repeated trials. The following theorem is key:

Theorem 15.5.1 (Central Limit Theorem) *Consider a sample space* $\omega = \{A, \text{not } A\}$ *with* $P(A) = p$ *and* $P(\text{not } A) = 1 - p = q$. *Given* n *independent random variables* $\mathbf{X}_1, \ldots, \mathbf{X}_n$, *each taking*

$$\mathbf{X}_i(A) = 1, \ \mathbf{X}_i(\text{not } A) = 0,$$

set

$$S_n = \sum_{i=1}^{n} \mathbf{X}_i$$

and

$$S_n^* = \frac{S_n - E(S_n)}{\sqrt{V(S_n)}}.$$

Then for any real numbers a *and* b,

$$\lim_{n \to \infty} P\{a \leq S_n^* \leq b\} = \frac{1}{\sqrt{2\pi}} \int_a^b e^{\frac{-x^2}{2}} \, \mathrm{d}x.$$

What this is saying is that if we perform a huge number of repeated Bernoulli trials, then the values of $\mathbf{S}_n$ will be distributed as:

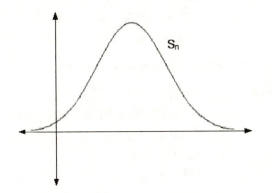

But we have even more. Namely, by normalizing S_n to the new random variable S_n^* (which, as we will see in a moment, has mean zero and variance one), we always get the same distribution, no matter what the real world situation we start with is, just as long as the real world problem can be modelled as a Bernoulli trial. By the way, the distribution for any Bernoulli trial is simply the graph of the function $\lim_{n\to\infty} S_n$. We call S_n^* the *normal distribution*. Its graph is the *Gauss-Bell curve*.

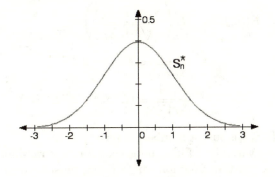

Before sketching a proof of the Central Limit Theorem (whose general outline is from [18]), let us look at the random variables S_n and S_n^*.

Lemma 15.5.1 *The expected value of S_n is np and its variance is npq. The expected value of S_n^* is 0 and its variance is 1.*

Proof of Lemma: We know that for all k,

$$E(\mathbf{X}_k) = \mathbf{X}_k(A)P(A) + \mathbf{X}_k(\text{not } A)P(\text{not } A) = 1 \cdot p + 0 \cdot q = p.$$

Then by the linearity of the expected value function

$$
\begin{aligned}
E(S_n) &= E(\mathbf{X}_1 + \ldots + \mathbf{X}_n) \\
&= E(\mathbf{X}_1) + \ldots + E(\mathbf{X}_n) \\
&= np.
\end{aligned}
$$

As for the variance, we know that for any k,

$$
\begin{aligned}
V(\mathbf{X}_k) &= E(\mathbf{X}_k^2) - [E(\mathbf{X}_k)]^2 \\
&= \mathbf{X}_k^2(A)P(A) + \mathbf{X}_k^2(\text{not } A)P(\text{not } A) - p^2 \\
&= 1^2 \cdot p + 0^2 \cdot q - p^2 \\
&= p - p^2 \\
&= p(1 - p) \\
&= pq.
\end{aligned}
$$

Then we have

$$
\begin{aligned}
V(S_n) &= V(\mathbf{X}_1 + \ldots + \mathbf{X}_n) \\
&= V(\mathbf{X}_1) + \ldots + V(\mathbf{X}_n) \\
&= npq.
\end{aligned}
$$

Now

$$
\begin{aligned}
E(S_n^*) &= E\left(\frac{S_n - E(S_n)}{\sqrt{V(S_n)}}\right) \\
&= \frac{1}{\sqrt{V(S_n)}} E(S_n - E(S_n)) \\
&= \frac{1}{\sqrt{V(S_n)}} (E(S_n) - E(E(S_n))),
\end{aligned}
$$

which, since $E(S_n)$ is just a number, is zero.

Now for the variance. First, note that for any random variable that happens to be a constant function, the variance must be zero. In particular, since the expected value of a random variable is a number, we must have that the variance of an expected value is zero:

$$
V(E(\mathbf{X})) = 0.
$$

Using this, we have that

$$
V(S_n^*) = V\left(\frac{S_n - E(S_n)}{\sqrt{V(S_n)}}\right)
$$

$$= (\frac{1}{\sqrt{V(S_n)}})^2 V(S_n - E(S_n))$$

$$= \frac{1}{V(S_n)}(V(S_n) - V(E(S_n)))$$

$$= 1,$$

as desired. $\square$

Before discussing the proof of the Central Limit Theorem, let us look at the formula

$$\lim_{n \to \infty} P(a \leq S_n^* \leq b) = \frac{1}{\sqrt{2\pi}} \int_a^b e^{\frac{-x^2}{2}} \, dx.$$

It happens to be the case that for any particular choice of a and b, it is impossible to explicitly calculate the integral $\frac{1}{\sqrt{2\pi}} \int_a^b e^{\frac{-x^2}{2}} dx$; instead people must numerically approximate the answers, which of course can easily be done with standard software packages like Maple or Mathematica. Surprisingly enough, $\frac{1}{\sqrt{2\pi}} \int_{-\infty}^{\infty} e^{\frac{-x^2}{2}} dx$ can be shown to be exactly one. We first show why this must be the case if the Central Limit Theorem is true and then we will explicitly prove that this integral is one.

For any sequence of events and for any n, S_n^* must be some number. Thus for all n,

$$P(-\infty \leq S_n^* \leq \infty) = 1,$$

and thus its limit as n goes to infinity must be one, meaning that our integral is one. Thus if $\frac{1}{\sqrt{2\pi}} \int_{-\infty}^{\infty} e^{\frac{-x^2}{2}} dx$ is not one, the Central Limit Theorem would not be true. Thus we need to prove that this integral is one. In fact, the proof that this integral is one is interesting in its own right.

Theorem 15.5.2

$$\frac{1}{\sqrt{2\pi}} \int_{-\infty}^{\infty} e^{\frac{-x^2}{2}} \, dx = 1.$$

Proof: Surprisingly, we look at the square of the integral:

$$(\frac{1}{\sqrt{2\pi}} \int_{-\infty}^{\infty} e^{\frac{-x^2}{2}} dx)^2 = (\frac{1}{\sqrt{2\pi}} \int_{-\infty}^{\infty} e^{\frac{-x^2}{2}} dx)(\frac{1}{\sqrt{2\pi}} \int_{-\infty}^{\infty} e^{\frac{-x^2}{2}} dx).$$

Since the symbol x just denotes what variable we are integrating over, we can change the x in the second integral to a y without changing the equality:

$$(\frac{1}{\sqrt{2\pi}} \int_{-\infty}^{\infty} e^{\frac{-x^2}{2}} dx)^2 = (\frac{1}{\sqrt{2\pi}} \int_{-\infty}^{\infty} e^{\frac{-x^2}{2}} dx)(\frac{1}{\sqrt{2\pi}} \int_{-\infty}^{\infty} e^{\frac{-y^2}{2}} dy).$$

Since the x and the y have nothing to do with each other, we can combine these two single integrals into one double integral:

$$(\frac{1}{\sqrt{2\pi}}\int_{-\infty}^{\infty}e^{\frac{-x^2}{2}}\mathrm{d}x)^2 \;=\; \frac{1}{2\pi}\int_{-\infty}^{\infty}\int_{-\infty}^{\infty}e^{\frac{-x^2}{2}}e^{\frac{-y^2}{2}}\mathrm{d}x\mathrm{d}y$$

$$=\; \frac{1}{2\pi}\int_{-\infty}^{\infty}\int_{-\infty}^{\infty}e^{\frac{-(x^2+y^2)}{2}}\mathrm{d}x\mathrm{d}y,$$

which is now a double integral over the real plane. The next trick is to switch over to polar coordinates, to reduce our integrals to doable ones. Recall that we have $\mathrm{d}x\mathrm{d}y = r\mathrm{d}r\mathrm{d}\theta$ and $x^2 + y^2 = r^2$

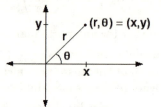

in polar coordinates. Then we have

$$(\frac{1}{\sqrt{2\pi}}\int_{-\infty}^{\infty}e^{\frac{-x^2}{2}}\mathrm{d}x)^2 \;=\; \frac{1}{2\pi}\int_{0}^{2\pi}\int_{0}^{\infty}-e^{-\frac{r^2}{2}}r\mathrm{d}r\mathrm{d}\theta$$

$$=\; \frac{1}{2\pi}\int_{0}^{2\pi}e^{-\frac{r^2}{2}}\,|_0^\infty\,\mathrm{d}\theta$$

$$=\; \frac{1}{2\pi}\int_{0}^{2\pi}\mathrm{d}\theta$$

$$=\; 1,$$

as desired. $\square$

Proof of Central Limit Theorem: (Again, we got this argument from [18].) At a critical stage of this proof, there will be a summation of terms of the form

$$\binom{n}{k}p^k q^{n-k}$$

which we will replace by

$$\frac{1}{\sqrt{2\pi npq}}e^{-\frac{x_k^2}{2}},$$

where the x_k will be defined in a moment. We will see that the justification for this replacement is a corollary of Stirling's formula for $n!$, next section's topic.

We are interested in $P(a \leq S_n^* \leq b)$. But, at least initially, the random variable S_n is a bit easier to work with. We want to link S_n with S_n^*. Suppose that we know that $S_n = k$, which means that after n trials, there have been exactly k occurrences of A (and thus $n-k$ occurrences of not A). Let x_k denote the corresponding value for S_n^*. Then

$$x_k = \frac{k - E(S_n)}{\sqrt{V(S_n)}}.$$

Since $E(S_n) = np$ and $V(S_n) = npq$, we have

$$x_k = \frac{k - np}{\sqrt{npq}},$$

and thus

$$k = np + \sqrt{npq}\,x_k.$$

Then

$$P(a \leq S_n^* \leq b) = \sum_{\{a \leq x_k \leq b\}} P(S_n = k).$$

First we need to show that

$$P(S_n = k) = \binom{n}{k} p^k q^{n-k}.$$

Now $S_n = k$ means that after n trials there are exactly k A's. Since $P(A) = p$ and $P(\text{not } A) = q$, we have that the probability of any particular pattern of k A's is $p^k q^{n-k}$ (for example, if the first k trials yield A's and the last $n - k$ trials yield not A's). But among n trials, there are $\binom{n}{k}$ different ways for there to be k A's. Thus $P(S_n = k) = \binom{n}{k} p^k q^{n-k}$.

Then we have

$$P(a \leq S_n^* \leq b) = \sum_{\{a \leq x_k \leq b\}} \binom{n}{k} p^k q^{n-k}.$$

We now replace $\binom{n}{k} p^k q^{n-k}$ with $\frac{1}{\sqrt{2\pi npq}} e^{-\frac{x_k^2}{2}}$ (which, again, will be justified in the next section), giving us

$$P(a \leq S_n^* \leq b) = \sum_{\{a \leq x_k \leq b\}} \frac{1}{\sqrt{2\pi npq}} e^{-\frac{x_k^2}{2}}$$

$$= \sum_{\{a \leq x_k \leq b\}} \frac{1}{\sqrt{2\pi}\sqrt{npq}} e^{-\frac{x_k^2}{2}}.$$

Note that
$$x_{k+1} - x_k = \frac{k+1-np}{\sqrt{npq}} - \frac{k-np}{\sqrt{npq}} = \frac{1}{\sqrt{npq}}.$$

Thus
$$P(a \le S_n^* \le b) = \sum_{\{a \le x_k \le b\}} \frac{1}{\sqrt{2\pi}} e^{\frac{x_k^2}{2}}(x_{k+1} - x_k).$$

As we let n approach infinity, the interval $[a, b]$ is split into a finer and finer partition by the x_k. The above sum is a Riemann sum and can thus be replaced, as n approaches infinity, by our desired integral:

$$\lim_{n \to \infty} P(a \le S_n^* \le b) = \frac{1}{\sqrt{2\pi}} \int_a^b e^{\frac{-x^2}{2}} \, \mathrm{d}x. \quad \square$$

15.6   Stirling's Approximation for $n!$

Stirling's formula tells us that for large n we can replace $n!$ by $\sqrt{2\pi n}n^n e^{-n}$. We need this approximation to complete the proof of the Central Limit Theorem. (We are still following [18].)

First, given two functions $f(n)$ and $g(n)$, we say that

$$f(n) \sim g(n)$$

if there exists a nonzero constant c such that

$$\lim_{n \to \infty} \frac{f(n)}{g(n)} = c.$$

Thus the functions $f(n)$ and $g(n)$ grow at the same rate as n goes to infinity. For example
$$n^3 \sim 5n^3 - 2n + 3.$$

Theorem 15.6.1 (Stirling's Formula)

$$n! \sim \sqrt{2\pi n}n^n e^{-n}$$

Proof: This will take some work and some algebraic manipulations.

First note that

$$\sqrt{2\pi n}n^n e^{-n} = \sqrt{2\pi}n^{n+\frac{1}{2}}e^{-n}.$$

We will show here that

$$\lim_{n \to \infty} \frac{n!}{n^{n+\frac{1}{2}}e^{-n}} = k,$$

for some constant k. To show that $k = \sqrt{2\pi}$, we use the following convoluted argument. Assume that we have already shown that $n! \sim kn^{n+\frac{1}{2}}e^{-n}$. Use this approximation in our replacement of $\binom{n}{k}p^k q^{n-k}$ in the following corollary and, more importantly, in the proof in the last section of the Central Limit Theorem. If we follow the steps in that proof, we will end up with

$$\lim_{n \to \infty} P(a \le S_n^* \le b) = \frac{1}{k} \int_a^b e^{\frac{-x^2}{2}} \, dx.$$

Since for each n, we must have S_n^* equal to some number, we know that $P(-\infty \le S_n^* \le \infty) = 1$ and thus $\lim_{n \to \infty} P(-\infty \le S_n^* \le \infty) = 1$. Then we must have

$$\frac{1}{k} \int_{-\infty}^{\infty} e^{\frac{-x^2}{2}} \, dx = 1.$$

But in the last section we calculated that $\int_{-\infty}^{\infty} e^{\frac{-x^2}{2}} \, dx = \sqrt{2\pi}$. From this calculation, we see that k must be $\sqrt{2\pi}$.

Now for the meat of the argument, showing that such a k exists. This will take some work and involve various computational tricks. Our goal is to show that there is a nonzero constant k such that

$$\lim_{n \to \infty} \frac{n!}{n^{n+\frac{1}{2}}e^{-n}} = k.$$

Since we have no clue for now as to what k is, save that it is positive, call it e^c, with c some other constant (we will be taking logarithms in a moment, so using e^c will make the notation a bit easier). Now,

$$\lim_{n \to \infty} \frac{n!}{n^{n+\frac{1}{2}}e^{-n}} = e^c$$

exactly when

$$\lim_{n \to \infty} \log\left(\frac{n!}{n^{n+\frac{1}{2}}e^{-n}}\right) = c.$$

Using that logarithms change multiplications and divisions into sums and differences, this is the same as

$$\lim_{n \to \infty} (\log(n!) - (n + \frac{1}{2})\log(n) + n) = c.$$

For notational convenience, set

$$d_n = \log(n!) - (n + \frac{1}{2})\log(n) + n.$$

We want to show that d_n converges to some number c as n goes to ∞. Here we use a trick. Consider the sequence

$$\sum_{i=1}^{n}(d_i - d_{i+1}) = (d_1 - d_2) + (d_2 - d_3) + \ldots (d_n - d_{n+1}) = d_1 - d_{n+1}.$$

We will show that the infinite series $\sum_{i=1}^{\infty}(d_i - d_{i+1})$ converges, which means that the partial sums $\sum_{i=1}^{n}(d_i - d_{i+1}) = d_1 - d_{n+1}$ converge. But this will mean that d_{n+1} will converge, which is our goal.

We will show that $\sum_{i=1}^{\infty}(d_i - d_{i+1})$ converges by the comparison test. Specifically, we will show that

$$|d_n - d_{n+1}| \leq \frac{2n+1}{2n^3} - \frac{1}{4n^2}.$$

Since both $\sum_{i=1}^{\infty}\frac{2n+1}{2n^3}$ and $\sum_{i=1}^{\infty}\frac{1}{4n^2}$ converge, our series will converge.

This will be a long calculation. We will need to use that, for any x with $|x| < \frac{2}{3}$,

$$\log(1 + x) = x - \frac{x^2}{2} + \theta(x)$$

where $\theta(x)$ is a function such that for all $|x| < \frac{2}{3}$,

$$|\theta(x)| < |x|^3.$$

This follows from the Taylor series expansion of $\log(1+x)$. The requirement that $|x| < \frac{2}{3}$ is not critical; all we must do is make sure that our $|x|$ are sufficiently less than one.

Now,

$$
\begin{aligned}
|d_n - d_{n+1}| &= [\log(n!) - (n + \tfrac{1}{2})\log(n) + n] - \\
&\quad [\log((n+1)!) - (n+1+\tfrac{1}{2})\log(n+1) + n + 1] \\
&= [\log(n) + \ldots + \log(1) - (n + \tfrac{1}{2})\log(n) + n] \\
&\quad -[\log(n+1) + \cdots + \log(1) \\
&\quad -(n+1+\tfrac{1}{2})\log(n+1) + n + 1] \\
&= -(n + \tfrac{1}{2})\log(n) + (n + \tfrac{1}{2})\log(n+1) - 1 \\
&= (n + \tfrac{1}{2})\log\left(\frac{n+1}{n}\right) - 1 \\
&= (n + \tfrac{1}{2})\log(1 + \tfrac{1}{n}) - 1
\end{aligned}
$$

$$= (n + \frac{1}{2})(\frac{1}{n} - \frac{1}{2n^2} + \theta(\frac{1}{n})) - 1$$

$$= (n + \frac{1}{2})\theta(\frac{1}{n}) - \frac{1}{4n^2}$$

$$\leq \frac{(n + \frac{1}{2})}{n^3} - \frac{1}{4n^2},$$

which gives us our result. $\square$

While Stirling's formula is important in its own right, we needed to use its following corollary in the proof of the Central Limit Theorem:

Corollary 15.6.1.1 *Let A be a constant. Then for $x_k \leq A$, we have*

$$\binom{n}{k} p^k q^{n-k} \sim \frac{1}{\sqrt{2\pi npq}} e^{-\frac{x_k^2}{2}}.$$

Here the notation is the same as that used in the last section. In particular, if $S_n = k$, we set $S_n^* = x_k$. Then we have

$$k = np + \sqrt{npq}x_k,$$

and subtracting both sides of this equation from n, we have

$$n - k = n - np - \sqrt{npq}x_k = nq - \sqrt{npq}x_k.$$

If, as in the corollary, $x_k \leq A$, then we must have

$$k \sim np$$

and

$$n - k \sim nq.$$

In the following proof, at a critical stage we will be replacing k by np and $n - k$ by nq.

Proof of Corollary: By definition

$$\binom{n}{k} p^k q^{n-k} = \frac{n!}{k!(n-k)!} p^k q^{n-k}$$

$$\sim \frac{(\frac{n}{e})^n \sqrt{2\pi n}}{(\frac{k}{e})^k \sqrt{2\pi k}(\frac{n-k}{e})^{n-k} \sqrt{2\pi(n-k)}} p^k q^{n-k},$$

using Stirling's formula, which in turn yields

$$= \sqrt{\frac{n}{2\pi k(n-k)}} \left(\frac{np}{k}\right)^k \left(\frac{nq}{n-k}\right)^{n-k}$$

$$\sim \sqrt{\frac{n}{2\pi(np)(nq)}} \left(\frac{np}{k}\right)^k \left(\frac{nq}{n-k}\right)^{n-k},$$

using here that $k \sim np$ and $n - k \sim nq$. This in turn equals

$$\sqrt{\frac{1}{2\pi npq}} \left(\frac{np}{k}\right)^k \left(\frac{nq}{n-k}\right)^{n-k}.$$

If we can show that

$$\left(\frac{np}{k}\right)^k \left(\frac{nq}{n-k}\right)^{n-k} \sim e^{-\frac{x_k^2}{2}},$$

we will be done. Using that we can replace $\log(1 + x)$ by $x - \frac{x^2}{2}$, for small x, we will show that

$$\log \left(\left(\frac{np}{k}\right)^k \left(\frac{nq}{n-k}\right)^{n-k} \right) \sim -\frac{x_k^2}{2}.$$

Now

$$\begin{aligned}
\log \left(\left(\frac{np}{k}\right)^k \left(\frac{nq}{n-k}\right)^{n-k} \right) &= k \log \left(\frac{np}{k}\right) + (n-k) \log \left(\frac{nq}{n-k}\right) \\
&= k \log \left(1 - \frac{\sqrt{npq}x_k}{k}\right) \\
&\quad + (n-k) \log \left(1 + \frac{\sqrt{npq}x_k}{n-k}\right),
\end{aligned}$$

using that the equality $k = np + \sqrt{npq}x_k$ implies

$$\frac{np}{k} = \frac{k - \sqrt{npq}x_k}{k} = 1 - \frac{\sqrt{npq}x_k}{k}$$

and a similar argument for the $(n - k)$. But then we can replace the log terms in the above to get

$$\begin{aligned}
&\sim \quad k \left(-\frac{\sqrt{npq}x_k}{k} - \frac{npqx_k^2}{2k^2}\right) + (n-k) \left(\frac{\sqrt{npq}x_k}{n-k} - \frac{npqx_k^2}{2(n-k)^2}\right) \\
&= \quad -\frac{npqx_k^2}{2k} - \frac{npqx_k^2}{2(n-k)} \\
&= \quad -\frac{npqx_k^2}{2} \left(\frac{1}{k} + \frac{1}{n-k}\right) \\
&= \quad -\frac{npqx_k^2}{2} \left(\frac{n}{k(n-k)}\right) \\
&= \quad -\frac{x_k^2}{2} \left(\frac{np}{k}\right) \left(\frac{nq}{n-k}\right) \\
&\sim \quad -\frac{x_k^2}{2},
\end{aligned}$$

since earlier we showed that $np \sim k$ and $nq \sim n - k$. $\square$

The proof of Stirling's formula and of its corollary were full of clever manipulations. Part of the reason that these steps are shown here is to let people see that despite the abstract machinery of modern mathematics, there is still a need for cleverness at computations.

15.7 Books

From informed sources, Brualdi's book [14] is a good introduction to combinatorics. An excellent, but hard, text is by van Lint and Wilson [115]. Cameron's text [16] is also good. Polya, Tarjan and Woods' book [93] is fascinating. To get a feel of how current combinatorics is used, Graham, Knuth and Patashnik's [47] book is great. Stanley's text [105] is a standard text for beginning graduate students in combinatorics.

For probability theory, it is hard to imagine a better text than Feller [34]. This book is full of intuitions and wonderful, nontrivial examples. Grimmett and Stirzaker [50] is also a good place to begin. Another good source is Chung's book [18], which is where, as mentioned, I got the flow of the above argument for the Central Limit Theorem. More advanced work in probability theory is measure theoretic.

15.8 Exercises

1. The goal of this exercise is to see how to apply the definitions for probability to playing cards.

a. Given a standard deck of fifty-two cards, how many five card hands are possible (here order does not matter).

b. How many of these five card hands contain a pair? (This means that not only must there be a pair in the hand, but there cannot be a three-of-a-kind, two pair, etc.)

c. What is the probability of being dealt a hand with a pair?

2. The goal of this exercise is to see how the formulas for $\binom{n}{k}$ are linked to Pascal's triangle.

a. Prove by induction that

$$\binom{n}{k} = \binom{n-1}{k} + \binom{n-1}{k-1}.$$

b. Prove this formula by counting how to choose k objects from n objects (order not mattering) in two different ways.

c. Prove that the binomial coefficients $\binom{n}{k}$ can be determined from Pascal's triangle, whose first five rows are:

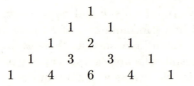

d. Give a combinatorial proof of the identity

$$\sum_{k=0}^{n} \binom{n}{k} = 2^n.$$

4. Find a formula for determining how many monomials of degree k can be made out of n variables. (Thus for the two variables x,y, the number of monomials of degree two is three, since we can simply count the list

$$(x^2, xy, y^2).)$$

5. The pigeonhole principle states:
If (n+1) objects are placed into n different boxes, at least one box must have at least two objects in it.

Let $a_1, \ldots, a_{n+1}$ be integers. Show that there is at least one pair of these integers such that $a_i - a_j$ is divisible by the integer n.

6. The goal of this problem is to prove the Inclusion-Exclusion Principle, the statement of which is part c.

a. Let A and B be any two sets. Show that

$$|A \cup B| = |A| + |B| - |A \cap B|.$$

b. Let A_1, A_2 and A_3 be any three sets. Show that

$$|A_1 \cup A_2 \cup A_3| = |A_1| + |A_2| + |A_3| - |A_1 \cap A_2| - |A_1 \cap A_3| - |A_2 \cap A_3| + |A_1 \cap A_2 \cap A_3|.$$

c. Let $A_1, \ldots, A_n$ be any n sets. Show that

$$|A_1 \cup \ldots \cup A_n| = \Sigma |A_i| - \Sigma |A_i \cap A_j| + \ldots + (-1)^{n+1} |A_1 \cap \ldots \cap A_n|.$$

7. Show that

$$\binom{2n}{n} \sim (\pi n)^{-1/2} 2^{2n}.$$

Chapter 16

Algorithms

Basic Object:	Graphs and Trees
Basic Goal:	Computing the Efficiency of Algorithms

The end of the 1800s and the beginning of the 1900s saw intense debate about the meaning of existence for mathematical objects. To some, a mathematical object could only have meaning if there was a method to compute it. For others, any definition that did not lead to a contradiction would be good enough to guarantee existence (and this is the path that mathematicians have overwhelmingly chosen to take). Think back to the section on the Axiom of Choice in Chapter Ten. Here objects were claimed to exist which were impossible to actually construct. In many ways these debates had quieted down by the 1930s, in part due to Gödel's work, but also in part due to the nature of the algorithms that were eventually being produced. By the late 1800s, the objects that were being supposedly constructed by algorithms were so cumbersome and time-consuming, that no human could ever compute them by hand. To most people, the pragmatic difference between an existence argument versus a computation that would take a human the life of the universe was too small to care about, especially if the existence proof had a clean feel.

All of this changed with the advent of computers. Suddenly, calculations that would take many lifetimes by hand could be easily completed in millionths of a second on a personal computer. Standard software packages like Mathematica and Maple can outcompute the wildest dreams of a mathematician from just a short time ago. Computers, though, seem to have problems with existence proofs. The need for constructive arguments returned with force, but now came a real concern with the efficiency of the construction, or the complexity of the algorithm. The idea that certain constructions have an intrinsic complexity has increasingly become basic in most branches of mathematics.

16.1   Algorithms and Complexity

An accurate, specific definition for an algorithm is non-trivial and not very enlightening. As stated in the beginning of Cormen, Leiserson and Rivest's book *Introduction to Algorithms* [22],

Informally, an **algorithm** *is any well-defined computational procedure that takes some value, or set of values, as* **input** *and produces some values, or set of values, as* **output**. *An algorithm is thus a sequence of computational steps that transform the input into the output.*

Much of what has been discussed in this book can be recast into the language of algorithms. Certainly, much of the first chapter on linear algebra, such as the definition of the determinant and Gaussian elimination, is fundamentally algorithmic in nature.

We are concerned with the efficiency of an algorithm. Here we need to be concerned with asymptotic bounds on the growth of functions.

Definition 16.1.1 *Let $f(x)$ and $g(x)$ be two one-variable real-valued functions. We say that $f(x)$ is in* $\mathrm{O(g(x))}$ *if there exists a positive constant C and a positive number N so that for all $x > N$, we have $|f(x)| \leq C|g(x)|$.*

This is informally known as *big O* notation.

Typically we do not use the symbol "x" for our variable but "n". Then the class of functions in $O(n)$ will be those that grow at most linearly, those in $O(n^2)$ grow at most quadratically, etc. Thus the polynomial $3n^4 + 7n - 19$ is in $O(n^4)$.

For an algorithm there is the *input size*, n, which is how much information needs to be initially given, and the *running time*, which is how long the algorithm takes as a function of the input size. An algorithm is *linear* if the running time $r(n)$ is in $O(n)$, *polynomial* if the running time $r(n)$ is in $O(n^k)$ for some integer k, etc.

There are further concerns, such as the space size of an algorithm, which is how much space the algorithm requires in order to run as a function of the input size.

16.2   Graphs: Euler and Hamiltonian Circuits

An analysis of most current algorithms frequently comes down to studying graphs. This section will define graphs and then discuss graphs that have Euler circuits and Hamiltonian circuits. We will see that while these two have similar looking definitions, their algorithmic properties are quite different.

Intuitively a graph looks like:

The key is that a graph consists of vertices and edges between vertices. All that matters is which vertices are linked by edges. Thus we will want these two graphs, which have different pictures in the plane, to be viewed as equivalent.

Definition 16.2.1 *A graph G consists of a set $V(g)$, called* vertices, *and a set $E(G)$, called* edges, *and a function*

$$\sigma : E(G) \to \{\{u,v\} : u, v \in V(G)\}.$$

We say that elements v_i and v_j in $V(G)$ are connected by an edge e if $\sigma(e) = \{v_i, v_j\}$.

Note that $\{v_i, v_j\}$ denotes the set consisting of the two vertices v_i and v_j.

For the graph G:

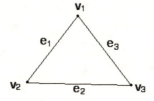

we have

$$V(G) = \{v_1, v_2, v_3\}$$
$$E(G) = \{e_1, e_2, e_3\}$$

and

$$\sigma(e_1) = \{v_1, v_2\}, \sigma(e_2) = \{v_2, v_3\}, \sigma(e_3) = \{v_1, v_3\}.$$

Associated to a graph is its *adjacency* matrix $A(G)$. If there are n vertices, this will be the following $n \times n$ matrix. List the vertices:

$$V(G) = \{v_1, v_2, ..., v_n\}.$$

For the (i, j)-entry of the matrix, put in a k if there are k edges between v_i and v_j and a 0 otherwise. Thus the adjacency matrix for:

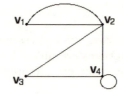

will be the 4×4 matrix:

$$A(G) = \begin{pmatrix} 0 & 2 & 0 & 0 \\ 2 & 0 & 1 & 1 \\ 0 & 1 & 0 & 1 \\ 0 & 1 & 1 & 1 \end{pmatrix}.$$

The '1' in the $(4, 4)$ entry reflects that there is an edge from v_4 to itself and the '2' in the $(1, 2)$ and $(2, 1)$ entries reflects that there are two edges from v_1 to v_2.

A *path* in a graph G is a sequence of edges that link up with each other. A *circuit* is a path that starts and ends at the same vertex. For example, in the graph:

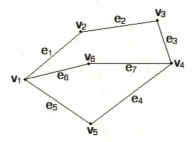

the path $e_6 e_7$ starts at vertex v_1 and ends at v_4 while $e_1 e_2 e_3 e_4 e_5$ is a circuit starting and ending at v_1.

We can now start to talk about Euler circuits. We will follow the traditional approach and look first of the Königsberg bridge problem. The town of Königsberg had the following arrangement:

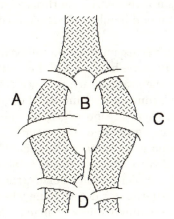

Here A, B, C and D denote land.

The story goes that in the 1700s, the people of Königsberg would try to see if they could cross every bridge exactly once so that at the end they returned to their starting spot. Euler translated this game into a graph theory question. To each connected piece of land he assigned a vertex and to each bridge between pieces of land he assigned an edge. Thus Königsberg became the graph

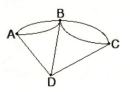

Then the game will be solved if in this graph there is a circuit that contains each edge exactly once. Such circuits have a special name, in honor of Euler:

Definition 16.2.2 *An* Euler circuit *on a graph is a circuit that contains each edge exactly once.*

To solve the Königsberg bridge problem, Euler came up with a clean criterion for when any graph will have an Euler circuit.

Theorem 16.2.1 *A graph has an Euler circuit if and only if each vertex has an even number of edges coming into it.*

Thus in Königsberg, since vertex A is on three edges (and in this case every other vertex also has an odd number of edges), no one can cross each bridge just once.

The fact that each vertex must be on an even number of edges is not that hard to see. Suppose we have an Euler circuit. Imagine deleting each edge as we transverse the graph. Each time we enter, then leave, a vertex, two edges are deleted, reducing the number of edges containing that vertex by two. By the end, there are no edges left, meaning that the original number of edges at each vertex had to be even.

The reverse direction is a bit more complicated but is more important. The best method (which we will not do) is to actually construct an algorithm that produces an Euler circuit. For us, the important point is that there is a clean, easy criterion for determining when an Euler circuit exists.

Let us now make a seemingly minor change in the definition for an Euler circuit. Instead of finding a circuit that contains each edge only once, now let us try to find one that contains each vertex only once. These circuits are called:

Definition 16.2.3 *A graph has a* Hamiltonian *circuit if there is a circuit that contains each vertex exactly once.*

For example, for the graph:

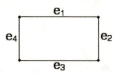

the circuit $e_1e_2e_3e_4$ is Hamiltonian, while for the graph:

there is no Hamiltonian circuit. In this last graph, one can simply list all possible circuits and then just check if one of them is Hamiltonian. This algorithm of just listing all possible circuits will work for any graph, as there can only be a finite number of circuits, but this listing unfortunately takes $O(n!)$ time, where n is the number of edges. For any graph with a fair number of edges, this approach is prohibitively time-consuming. But this is fairly close to the best known method for determining if a Hamiltonian circuit exists. As we will see in section four, the problem of finding a Hamiltonian circuit seems to be intrinsically difficult and important.

16.3 Sorting and Trees

Suppose you are given a set of real numbers. Frequently you want to order the set from smallest number to largest. Similarly, suppose a stack of exams is sitting on your desk. You might want to put the exams into alphabetical order. Both of these problems are sorting problems. A sorting algorithm will take a collection of elements for which an ordering can exist and actually produce the ordering. This section will discuss how this is related to a special class of graphs called *trees* and that the lower bound for any sorting algorithm is $O(n \log(n))$.

Technically a *tree* is any graph that is connected (meaning that there is a path from any vertex to any other vertex) and contains within it no circuits. Thus

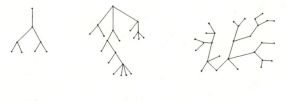

are trees while

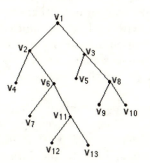

are not. Those vertices contained on exactly one edge are called *leaves*. These are in some sense the vertices where the tree stops. We will be concerned with *binary* trees, which are constructed as follows. Start with a vertex called the *root*. Let two edges come out from the root. From each of the two new vertices at the end of the two edges, either let two new edges stem out or stop. Continue this process a finite number of steps. Such a tree looks like:

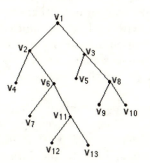

where v_1 is the root and v_4, v_5, v_7, v_9, v_{10}, v_{12} and v_{13} are the leaves. We
will draw our binary trees top down, with the root at the top and the leaves
at the bottom. At each vertex, the two edges that stem down are called
the *left* edge and *right* edge, respectively. The two vertices at the ends of
these edges are called the *left* child and the *right* child, respectively. The
height of a tree is the number of edges in the longest path from the root to
a leaf. Thus the height of

is three while the height of

is six.

We now want to see why sorting is linked to binary trees. We are given
a collection of elements $\{a_1, \ldots, a_n\}$. We will assume that all we can do is
compare the size of any two elements. Thus given, say, elements a_i and a_j,
we can determine if $a_i < a_j$ or if $a_j < a_i$. Any such sorting algorithm can
only, at each stage, take two a_i and a_j and, based on which is larger, tell
us what to do at the next stage. Now to show that any such algorithm can
be represented as a tree. The root will correspond to the first pair to be
compared in the algorithm. Say this first pair is a_i and a_j. There are two
possibilities for the order of a_i and a_j. If $a_i < a_j$, go down the left edge
and if $a_j < a_i$, go down the right edge. An algorithm will tell us at this
stage which pair of elements to now compare. Label the new vertices by
these pairs. Continue this process until there is nothing left to compare.
Thus we will have a tree, with each vertex labeled by a pair of elements in
our set and each leaf corresponding to an ordering of the set.

For example, take a three element set $\{a_1, a_2, a_3\}$. Consider the following simple algorithm (if anything this easy deserves to be called an
algorithm):
Compare a_1 and a_2. If $a_1 < a_2$, compare a_2 and a_3. If $a_2 < a_3$, then the
ordering is $a_1 < a_2 < a_3$. If $a_3 < a_2$, compare a_1 and a_3. If $a_1 < a_3$,
then the ordering is $a_1 < a_3 < a_2$. If we had $a_3 < a_1$, then the ordering
is $a_3 < a_1 < a_2$. Now we go back to the case when $a_2 < a_1$. Then we

next compare a_1 and a_3. If $a_1 < a_3$, the ordering is $a_2 < a_1 < a_3$. If
we have $a_3 < a_1$, we compare a_2 and a_3. If $a_2 < a_3$, then the ordering is
$a_2 < a_3 < a_1$. If $a_3 < a_2$, then the ordering is $a_3 < a_2 < a_1$ and we are
done. Even for this simple example, the steps, presented in this manner,
are confusing. But when this method is represented as a tree it becomes
clear:

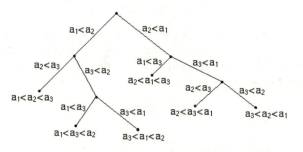

We now want to show that for a binary tree there is an intrinsic lower
bound on its height, which means that there is an intrinsic lower bound on
the time needed to sort.

Theorem 16.3.1 *A binary tree of height n has at most 2^n leaves.*

Proof: By induction. Suppose the height is zero. This means that the tree
is a single vertex and thus has $2^0 = 1$ leaf, which of course in this case is
also the root and is easy to sort.

Now suppose that we know the theorem is true for any tree of height
$n-1$. Look at a tree of height n. Thus there is at least one path from
the root to a leaf with length n. Remove all leaves, and their attaching
edges, that are of length n from the root. We have a new tree of height
$n-1$. The induction hypothesis kicks in, so we know that for this new tree
there are at most 2^{n-1} leaves. Let two edges stem out from each of these
2^{n-1} leaves, forming still another new tree which has height n and which
contains our original tree. But we are adding two new vertices for each of
the 2^{n-1} leaves of the tree of height $n-1$. Thus this final new tree has at
most $2 \cdot 2^{n-1} = 2^n$ leaves. Since each leaf of our original tree is a leaf of
this tree, we have our result. $\square$

This allows us to finally see that any algorithm that sorts n objects
must be in at least $O(n \log(n))$.

Theorem 16.3.2 *Any sorting algorithm based on pairwise comparisons
must be in at least $O(n \log(n))$.*

Proof: Given a set of n elements, there are $n!$ different ways they can be
initially ordered. For any sorting algorithm, for the corresponding tree there

must be a way, starting with the root, to get to one of these $n!$ different initial orderings. Thus the tree must have at least $n!$ leaves. Thus from the previous theorem, the tree must have height at least h, where

$$2^h \geq n!.$$

Thus we must have

$$h \geq \log_2(n!).$$

Any sorting algorithm must take at least h steps and hence must be in at least $O(\log_2(n!))$. Now we have, for any number K, $\log(K) = \log(2)\log_2(K)$, where of course, log is here the natural log, $\log_e$. Further, by Stirling's formula, we have for large n that

$$n! \sim \sqrt{2\pi n}\, n^n e^{-n}.$$

Then

$$\log(n!) \sim \log(\sqrt{2\pi n}) + n\log(n) - n\log(e),$$

which gives us that

$$
\begin{aligned}
O(\log(n!)) &= O(\log(\sqrt{2\pi n}) + n\log(n) - n\log(e)) \\
&= O(n\log(n)),
\end{aligned}
$$

since $n\log(n)$ dominates the other terms. Thus the complexity of any sorting algorithm is in at least $O(\log_2(n!)$, which equals $O(n\log(n))$, as desired. $\square$

To show that sorting is actually equal to $O(n\log(n))$, we would need to find an algorithm that runs in $O(n\log(n))$. Heapsort, merge and other algorithms for sorting do exist that are in $O(n\log(n))$.

16.4   P=NP?

The goal of this section is to discuss what is possibly the most important open problem in mathematics: "P=NP?". This problem focuses on trying to determine the difference between the finding of a solution for a problem and the checking of a candidate solution for the problem. The fact that it remains open (and that it could well be independent of the other axioms of mathematics) shows that mathematicians do not yet understand the full meaning of mathematical existence versus construction.

A problem is in *polynomial* time if, given input size n, there is an algorithm that is in $O(n^k)$, for some positive integer k. A problem is in NP if, given input size n, a candidate solution can be checked for accuracy in polynomial time. The N in the NP is somewhat of a joke; NP stands for "not polynomial".

Think of a jigsaw puzzle. While it can be quite time consuming to put a jigsaw puzzle together, it is easy and quick to tell if someone has finished such a puzzle. For a more mathematical example, try to invert an $n \times n$ matrix A. While doable, it is not particularly easy to actually construct A^{-1}. But if someone hands us a matrix B and claims that it is the inverse, all we have to do to check is to multiply out AB and see if we get the identity I. For another example, start with a graph G. It is difficult to determine if G contains a Hamiltonian circuit. But if someone hands us a candidate circuit, it is easy to check whether or not the circuit goes through every vertex exactly once. Certainly it appears that the problem of finding a solution should be intrinsically more difficult than the problem of checking the accuracy of a solution.

Amazingly enough, people do not know if the class of NP problems is larger than the class of polynomial time problems (which are denoted as P problems). "P=NP" is the question:

Is the class of problems in P equal to the class of problems in NP?

This has been open for many years. While initially the smart money was on P≠NP, today the belief is increasingly that statement 'P=NP' is independent of the other axioms of mathematics. Few believe that P=NP.

Even more intriguing is the existence of NP *complete* problems. Such a problem is not only in NP but also must be a yes/no question and, most importantly, every other NP problem must be capable of being translated into this problem in polynomial time. Thus if there is a polynomial time solution to this NP yes/no problem, there will be a polynomial time solution of every NP problem.

Every area of math seems to have its own NP complete problems. For example, the question of whether or not a graph contains a Hamiltonian circuit is a quintessential NP complete problem and, since it can be explained with little high level math, is a popular choice in expository works.

16.5 Numerical Analysis: Newton's Method

Since the discovery of calculus, there has been work on finding answers to math questions that people can actually use. Frequently this comes down to only finding approximate solutions. Numerical Analysis is the field that tries to find approximate solutions to exact problems. How good of an approximation is good enough and how quickly the approximation can be found are the basic questions for a numerical analyst. While the roots of this subject are centuries old, the rise of computers has revolutionized the field. An algorithm that is unreasonable to perform by hand can often be easily solved for a standard computer. Since numerical analysis is ultimately concerned with the efficiency of algorithms, I have put this section in this chapter. It must be noted that in the current math world, numerical analysts and people in complexity theory are not viewed as being in the same subdiscipline. This is not to imply that they don't talk to each other; more that complexity theory has evolved from computer science and numerical analysis has always been a part of mathematics.

There are certain touchstone problems in numerical analysis, problems that are returned to again and again. Certainly efficient algorithms for computations in linear algebra are always important. Another, which we will be concerned with here, is the problem of finding zeros of functions. Many problems in math can be recast into finding a zero of a function. We will first look at Newton's method for approximating a zero of a real valued differentiable function $f : \mathbf{R} \to \mathbf{R}$, and then quickly see how the ideas behind this method can be used, at times, to approximate the zeros of other types of functions.

Let $f : \mathbf{R} \to \mathbf{R}$ be a differentiable function. We will first outline the geometry behind Newton's method. Suppose we know its graph (which of course in real life we will rarely know; otherwise the problem of approximating zeros would be easy) to be:

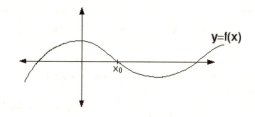

We thus want to approximate the point x_0. Choose any point x_1. Draw the tangent line to the curve $y = f(x)$ at the point $(x_1, f(x_1))$ and label its intersection with the x-axis by $(x_2, 0)$.

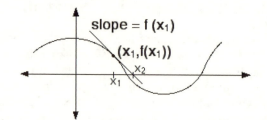

Then we have

$$f'(x_1) = \frac{0 - f(x_1)}{x_2 - x_1},$$

which, solving for x_2, yields

$$x_2 = x_1 - \frac{f(x_1)}{f'(x_1)}.$$

In the picture, it looks like our newly constructed x_2 is closer to our desired x_0 than is x_1. Let us try the same thing but replacing the x_1's with x_2. We label x_3 as the x-coordinate of the point of intersection of the tangent line of $y = f(x)$ through the point $(x_2, f(x_2))$ and get:

$$x_3 = x_2 - \frac{f(x_2)}{f'(x_2)}.$$

Again, it at least looks like x_3 is getting closer to x_0. Newton's method is to continue this process, namely to set

$$x_{k+1} = x_k - \frac{f(x_k)}{f'(x_k)}.$$

For this to work, we need $x_k \to x_0$. There are difficulties. Consider the picture:

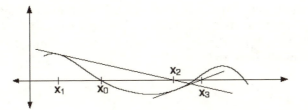

With this choice of initial x_1, the x_k will certainly not approach the zero x_0, though they do appear to approach a different zero. The problem of course is that this choice of x_1 is near a local maximum, which means that the derivative $f'(x_1)$ is very small, forcing $x_2 = x_1 - f(x_1)/f'(x_1)$ to be far from x_0.

We will now make this technically correct. Here we will see many ideas from calculus playing a critical role in proving that Newton's method will, subject to specific conditions, always produce an approximation to the true zero. We will look at functions $f : [a, b] \to [a, b]$ which have continuous second derivatives, i.e., functions in the vector space $C^2[a, b]$. As an aside, we will be using throughout the Mean Value Theorem, which states that for any function $f \in C^2[a, b]$, there exists a number c with $a \leq c \leq b$ such that

$$f'(c) = \frac{f(b) - f(a)}{b - a}.$$

Our goal is:

Theorem 16.5.1 *Let $f \in C^2[a, b]$. Suppose there exists a point $x_0 \in [a, b]$ with $f(x_0) = 0$ but $f'(x_0) \neq 0$. Then there exists a $\delta > 0$ such that, given any point x_1 in $[x_0 - \delta, x_0 + \delta]$, if for all k we define*

$$x_k = x_{k-1} - \frac{f(x_{k-1})}{f'(x_{k-1})},$$

we have that $x_k \to x_0$.

This theorem states that Newton's method will produce an approximation of the zero provided our initial choice x_1 is close enough to the zero.
Proof: We will alter the problem from finding a zero of a function f to the finding of a fixed point of a function g. Set

$$g(x) = x - \frac{f(x)}{f'(x)}.$$

Note that $f(x_0) = 0$ if and only if $g(x_0) = x_0$. We will show that Newton's method will produce an approximation to a fixed point of g.

We first need to see how to choose our $\delta > 0$. By taking derivatives and doing a bit of algebra, we have

$$g'(x) = \frac{f(x) f''(x)}{(f'(x))^2}.$$

Since the second derivative of f is still a continuous function, we have that $g'(x)$ is a continuous function. Further, since $f(x_0) = 0$, we have that $g'(x_0) = 0$. By continuity, given any positive number α, there exists a $\delta > 0$ such that for all $x \in [x_0 - \delta, x_0 + \delta]$, we have

$$|g'(x)| < \alpha.$$

We choose α to be strictly less than one (the reason for this restriction will be clear in a moment).

We will reduce the problem to proving the following three lemmas:

Lemma 16.5.1 *Let $g : [a, b] \to [a, b]$ be any continuous function. Then there is a fixed point in $[a, b]$.*

Lemma 16.5.2 *Let $g : [a, b] \to [a, b]$ be any differentiable function such that for all $x \in [a, b]$ we have*

$$|g'(x)| < \alpha < 1$$

for some constant α. Then there is a unique fixed point in the interval $[a, b]$.

Lemma 16.5.3 *Let $g : [a, b] \to [a, b]$ be any differentiable function such that for all $x \in [a, b]$ we have*

$$|g'(x)| < \alpha < 1$$

for some constant α. Then given any $x_1 \in [a, b]$, if we set

$$x_{k+1} = g(x_k),$$

then the x_k will approach the fixed point of g.

Assume briefly that all three lemmas are true. Note by our choice of δ, we have the function $g(x) = x - \frac{f(x)}{f'(x)}$ satisfying each of the conditions in the above lemma. Further we know that the zero x_0 of the function $f(x)$ is the fixed point of $g(x)$. Then we know that iterating any point in $[x_0 - \delta, x_0 + \delta]$ by g, we will approach x_0. But writing out this iteration is precisely Newton's method.

Now to prove the lemmas.

Proof of first lemma: This will be a simple application of the Intermediate Value Theorem. If $g(a) = a$ or if $g(b) = b$, then a or b is our fixed point and we are done. Suppose neither holds. Since the range of g is contained in the interval $[a, b]$, this means that

$$a < g(a) \text{ and } b > g(b).$$

Set

$$h(x) = x - g(x).$$

This new function is continuous and has the property that

$$h(a) = a - g(a) < 0$$

and

$$h(b) = b - g(b) > 0.$$

By the Intermediate Value Theorem, there must be a $c \in [a, b]$ with

$$h(c) = c - g(c) = 0$$

giving us our fixed point. $\square$

Proof of second lemma: We will now use the Mean Value Theorem. Suppose there are two distinct fixed points, c_1 and c_2. Label these points so that $c_1 < c_2$. By the Mean Value Theorem, there is some number c with $c_1 \leq c \leq c_2$ such that

$$\frac{g(c_2) - g(c_1)}{c_2 - c_1} = g'(c).$$

Since $g(c_1) = c_1$ and $g(c_2) = c_2$, we have

$$g'(c) = \frac{c_2 - c_1}{c_2 - c_1} = 1.$$

Here is our contradiction, as we assumed that at all points that the absolute value of the derivative was strictly less than one. There cannot be two fixed points. $\square$

Proof of third lemma: This will be another application of the Mean Value Theorem. By the second lemma, we know that g has a unique fixed point. Call this fixed point x_0. We will regularly replace x_0 by $g(x_0)$.

Our goal is to show that $|x_k - x_0| \to 0$. We will show that for all k

$$|x_k - x_0| \leq \alpha |x_{k-1} - x_0|.$$

Then by shifting subscripts we will have

$$|x_{k-1} - x_0| \leq \alpha |x_{k-2} - x_0|,$$

which will mean that

$$|x_k - x_0| \leq \alpha |x_{k-1} - x_0| \leq \alpha^2 |x_{k-2} - x_0| \leq \ldots \leq \alpha^k |x_1 - x_0|.$$

Since α is strictly less than one, we will have $|x_k - x_0| \to 0$.

Now

$$|x_k - x_0| = |g(x_{k-1}) - g(x_0)|.$$

By the Mean Value Theorem, there is some point c between x_0 and x_{k-1} with

$$\frac{g(x_{k-1}) - g(x_0)}{x_{k-1} - x_0} = g'(c)$$

which is equivalent to

$$g(x_{k-1}) - g(x_0) = g'(c)(x_{k-1} - x_0).$$

Then

$$|g(x_{k-1}) - g(x_0)| = |g'(c)||x_{k-1} - x_0|.$$

Now we just have to observe that by assumption $|g'(c)| \leq \alpha$, and we are done. $\square$

All this theorem is telling us is that if we start with an initial point close enough to the zero of a function, Newton's method will indeed converge to the zero. It does not tell us how to make our initial choice and does not tell us the speed of the convergence.

Now let us see how to try to use Newton's method in other contexts. Suppose we have a map $L : V \to W$ from one vector space to another. How can we approximate a zero of this map? Let us assume that there is some notion of a derivative for the map L, which we will denote by DL. Then just formally following the Newton's method, we might, starting with any element $v_1 \in V$, recursively define

$$v_{k+1} = v_k - DL(v_k)^{-1} L(v_k)$$

and hope that the v_k will approach the zero of the map. This could be at least an outline of a general approach. The difficulties are in understanding DL and in particular in dealing with when DL has some type of inverse.

For example, consider a function $F : \mathbf{R}^2 \to \mathbf{R}^2$, given in local coordinates by

$$F(x, y) = (f_1(x, y), f_2(x, y)).$$

The derivative of F should be the two-by-two Jacobian matrix

$$DF = \begin{pmatrix} \frac{\partial f_1}{\partial x} & \frac{\partial f_1}{\partial y} \\ \frac{\partial f_2}{\partial x} & \frac{\partial f_2}{\partial y} \end{pmatrix}.$$

Starting with any $(x_1, y_1) \in \mathbf{R}^2$, we set

$$\begin{pmatrix} x_{k+1} \\ y_{k+1} \end{pmatrix} = \begin{pmatrix} x_k \\ y_k \end{pmatrix} - DF^{-1}(x_k, y_k) \cdot \begin{pmatrix} f_1(x_k, y_k) \\ f_2(x_k, y_k) \end{pmatrix}.$$

Newton's method will work if the (x_k, y_k) approach a zero of F. By placing appropriate restrictions on the zero of F, such as requiring that $\det(DF(x_0, y_0)) \neq 0$, we can find an analogous proof to the one-dimensional case. In fact, it generalizes to any finite dimension.

More difficult problems occur for infinite dimensional spaces V and W. These naturally show up in the study of differential equations. People still try to follow a Newton-type method, but now the difficulty of dealing with the right notion for DL becomes a major stumbling block. This is why in trying to solve differential equations you are led to the study of infinite dimensional linear maps and are concerned with the behavior of the eigenvalues, since you want to control and understand what happens when the eigenvalues are, or are close to, zero, for this is the key to controlling the inverse of DL. The study of such eigenvalue questions falls under the rubric of Spectral Theorems, which is why the Spectral Theorem is a major part of beginning Functional Analysis and a major tool in PDE theory.

16.6   Books

The basic text for algorithms is *Introduction to Algorithms* by Cormen, Leiserson and Rivest [22]. Another source is *Data Structures and Algorithms* by Aho, Hopcroft, and Ullman [2].

Numerical Analysis has a long history. Further, many people, with widely varying mathematical backgrounds, need to learn some numerical analysis. Thus there are many beginning texts (though it must be stated that my knowledge of these texts is limited).

Atkinson's *Introduction to Numerical Analysis* [5] comes highly recommended. Another basic text that has long been the main reference for people studying for the numerical methods part of the actuarial exams is *Numerical Methods* by Burden and Faires [15]. Trefethon and Bau's text [112] is a good source for numerical methods for linear algebra. For numerical methods for differential equations, good sources are the books by Iserles [66] and Strikwerda [110]. Finally, for links with optimization theory, there is Ciarlet's *Introduction to Numerical Linear Algebra and Optimization* [19].

16.7   Exercises

1. Show that there are infinitely many nonisomorphic graphs, each having exactly k vertices.

2. How many nonisomorphic graphs with exactly three vertices and four edges are there?

3. Assume that the time for multiplying and adding two numbers together is exactly one.

   a. Find an algorithm that runs in time (n-1) that adds n numbers together.

   b. Find an algorithm that computes the dot product of two vectors in $\mathbf{R}^2$ in time (2n-1).

   c. Assume that we can work in parallel, meaning that we allow algorithms that can compute items that do not depend on each other simultaneously. Show that we can add n numbers together in time $\log_2(n-1)$.

   d. Find an algorithm that computes the dot product of two vectors in $\mathbf{R}^2$ in parallel in time $\log_2(n)$.

4. Let A be the adjacency matrix for a graph G.

   a. Show that there is a nonzero (i, j) entry of the matrix A^2 if and only if there is a path containing two edges from vertex i to vertex j.

   b. Generalize part (a) to linking entries in the matrix A^k to the existence of paths between various vertices having exactly k edges.

c. Find an algorithm that determines whether or not a given graph is connected.

5. Use Newton's method, with a calculator, to approximate $\sqrt{2}$ by approximating a root of the polynomial $x^2 - 2$.

6. Let $f : \mathbf{R}^n \to \mathbf{R}^n$ be any differentiable function from $\mathbf{R}^n$ to itself. Let x_0 be a point in $\mathbf{R}^n$ with $f(x_0) = 0$ but with $\det(Df(x_0)) \neq 0$, where Df denotes the Jacobian of the function f. Find a function $g : \mathbf{R}^n \to \mathbf{R}^n$ that has the point x_0 as a fixed point.

Appendix A

Equivalence Relations

Throughout this text we have used equivalence relations. Here we collect some of the basic facts about equivalence relations. In essence, an equivalence relation is a generalization of equality.

Definition A.0.1 (Equivalence Relation) *An equivalence relation on a set X is any relation '$x \sim y$' for $x, y \in X$ such that*

1. *(Reflexivity) For any $x \in X$, we have $x \sim x$.*

2. *(Symmetry) For all $x, y \in X$, if $x \sim y$ then $y \sim x$.*

3. *(Transitivity) For all $x, y, z \in X$, if $x \sim y$ and $y \sim z$, then $x \sim z$.*

The basic example is that of equality. Another example would be when $X = \mathbf{R}$ and we say that $x \sim y$ if $x - y$ is an integer. On the other hand, the relation $x \sim y$ if $x \leq y$ is not an equivalence relation, as it is not symmetric.

We can also define equivalence relations in term of subsets of the ordered pairs $X \times X$ as follows:

Definition A.0.2 (Equivalence Relation) *An equivalence relation on a set X is a subset $R \subset X \times X$ such that*

1. *(Reflexivity) For any $x \in X$, we have $(x, x) \in R$.*

2. *(Symmetry) For all $x, y \in X$, if $(x, y) \in R$ then $(y, x) \in R$.*

3. *(Transitivity) For all $x, y, z \in X$, if $(x, y) \in R$ and $(y, z) \in R$, then $(x, z) \in R$.*

The link between the two definitions is of course that $x \sim y$ means the same as $(x, y) \in R$.

An equivalence relation will split the set X into disjoint subsets, the equivalence classes.

Definition A.0.3 (Equivalence Classes) *An* equivalence class C *is a subset of X such that if $x, y \in C$, then $x \sim y$ and if $x \in C$ and if $x \sim y$, then $y \in C$.*

The various equivalence classes are disjoint, a fact that follows from transitivity.

Exercises: 1. Let G be a group and H a subgroup. Define, for $x, y \in G$, $x \sim y$, whenever $xy^{-1} \in H$. Show that this forms an equivalence relation on the group G.

2. For any two sets A and B, define $A \sim B$ if there is a one-to-one, onto map from A to B. Show that this is an equivalence relation.

3. Let (v_1, v_2, v_3) and (w_1, w_2, w_3) be two collections of three vectors in $\mathbf{R}^3$. Define $(v_1, v_2, v_3) \sim (w_1, w_2, w_3)$ if there is an element $A \in \mathbf{GL(n,R)}$ such that $Av_1 = w_1$, $Av_2 = w_2$ and $Av_3 = w_3$. Show that this is an equivalence relation.

4. On the real numbers, say that $x \sim y$ if $x - y$ is a rational number. Show that this forms an equivalence relation on the real numbers. (This equivalence was used in Chapter Ten, in the proof that there exists non-measurable sets.)

Bibliography

[1] Ahlfors, Lars V., *Complex Analysis: An Introduction to the Theory of Analytic Functions of One Complex Variable*, Third edition, International Series in Pure and Applied Mathematics, McGraw-Hill Book Co., New York, 1978. xi+331 pp.

[2] Aho, A., Hopcroft, J. and Ullman, J., *The Design and Analysis of Computer Algorithms*, Addison-Wesley, Reading, NY, 1974.

[3] Artin, E., *Galois Theory*, (edited and with a supplemental chapter by Arthur N. Milgram), Reprint of the 1944 second edition, Dover Publications, Inc., Mineola, NY, 1998. iv+82 pp.

[4] Artin, M., *Algebra*, Prentice Hall, 1995. 672 pp.

[5] Atkinson, Kendall E., *An Introduction to Numerical Analysis*, Second edition, John Wiley and Sons, Inc., New York, 1989. xvi+693 pp.

[6] Bartle, Robert G., *The Elements of Real Analysis*, Second edition, John Wiley and Sons, New York-London-Sydney, 1976. xv+480 pp.

[7] Berberian, Sterling K., *A First Course in Real Analysis*, Undergraduate Texts in Mathematics, Springer-Verlag, New York, 1998. xii+237 pp.

[8] Berenstein, Carlos A.and Gay, Roger, *Complex Variables: An Introduction*, Graduate Texts in Mathematics, 125. Springer-Verlag, New York, 1997. xii+650 pp.

[9] Birkhoff, G. and Mac Lane, S, *A Survey of Modern Algebra*, Akp Classics, A K Peters Ltd, 1997. 512 pp.

[10] Bocher, M., *Introduction to Higher Algebra*, MacMillan, New York, 1907.

[11] Bollobas, B., *Graph Theory: An Introductory Course*, Graduate Texts in Mathematics, 63, Springer-Verlag, New York-Berlin, 1979. x+180 pp.

[12] Boyce, W.F. and Diprima, R. C., *Elementary Differential Equations and Boundary Value Problems*, Sixth Edition, John Wiley and Sons, 1996, 768 pp.

[13] Bressoud, David M., *A Radical Approach to Real Analysis* Classroom Resource Materials Series, 2, Mathematical Association of America, Washington, DC, 1994. xii+324 pp.

[14] Brualdi, Richard A., *Introductory Combinatorics*, Second edition, North-Holland Publishing Co., New York, 1992. xiv+618 pp.

[15] Burden, R. and Faires, J., *Numerical Methods*, Seventh edition, Brooks/Cole Publishing Co., Pacific Grove, CA, 2001. 810 pp

[16] Cameron, Peter J., *Combinatorics: Topics, Techniques, Algorithms*, Cambridge University Press, Cambridge, 1995. x+355 pp.

[17] Cederberg, Judith N, *A Course in Modern Geometries*, Second edition, Undergraduate Texts in Mathematics, Springer-Verlag, New York-Berlin, 2001. xix+439 pp.

[18] Chung, Kai Lai, *Elementary Probability Theory with Stochastic Processes*, Second printing of the second edition, Undergraduate Texts in Mathematics. Springer-Verlag New York, New York-Heidelberg, 1975. x+325 pp.

[19] Ciarlet, Phillippe,, *Introduction to Numerical Linear Algebra and Optimisation* Cambridge Texts in Applied Mathematics, Vol. 2, Cambridge University Press, 1989, 452 pp.

[20] Cohen, Paul J., *Set Theory and the Continuum Hypothesis*, W. A. Benjamin, Inc., New York-Amsterdam 1966 vi+154 pp.

[21] Conway, John B., *Functions of One Complex Variable* Second edition. Graduate Texts in Mathematics, 11, Springer-Verlag, New York-Berlin, 1995. xiii+317 pp.

[22] Cormen, Thomas H.; Leiserson, Charles E.; Rivest, Ronald L., *Introduction to Algorithms*, The MIT Electrical Engineering and Computer Science Series, MIT Press, Cambridge, MA; McGraw-Hill Book Co., New York, 1990. xx+1028 pp.

[23] Coxeter, H. S. M., *Introduction to Geometry*, Second edition, Reprint of the 1969 edition, Wiley Classics Library. John Wiley and Sons, Inc., New York, 1989. xxii+469 pp.

[24] Davis, Harry, *Fourier Series and Orthogonal Functions*, Dover, 1989, 403 pp.

[25] Davis, Philip J., *The Schwarz function and its applications*, The Carus Mathematical Monographs, No. 17, The Mathematical Association of America, Buffalo, N. Y., 1974. 241 pp.

[26] De Souza, P. and Silva, J., *Berkeley Problems in Mathematics*, Springer-Verlag, New York, 1998, 457 pp.

[27] do Carmo, Manfredo P., *Differential Forms and Applications*, Universitext, Springer-Verlag, Berlin, 1994. x+118 pp.

[28] do Carmo, Manfredo P., *Riemannian Geometry*, Translated from the second Portuguese edition by Francis Flaherty, Mathematics: Theory & Applications, Birkhauser Boston, Inc., Boston, MA, 1994. xiv+300 pp.

[29] do Carmo, Manfredo P., *Differential Geometry of Curves and Surfaces*, Prentice-Hall, Inc., Englewood Cliffs, N.J., 1976. viii+503 pp.

[30] Dugundji, James, *Topology*, Allyn and Bacon, Inc., Boston, Mass. 1966 xvi+447 pp.

[31] Edwards, H., *Galois Theory*, Graduate Texts in Mathematics, 101, Springer, 1984.

[32] Euclid, *The Thirteen Books of Euclid's Elements*, translated from the text of Heiberg. Vol. I: Introduction and Books I, II., Vol. II: Books III–IX, Vol. III: Books X–XIII and Appendix, Translated with introduction and commentary by Thomas L. Heath, Second edition, Dover Publications, Inc., New York, 1956. xi+432 pp.; i+436 pp.; i+546 pp.

[33] Evans, Lawrence C., *Partial Differential Equations*, Graduate Studies in Mathematics, 19, American Mathematical Society, Providence, RI, 1998. xviii+662 pp.

[34] Feller, William, *An Introduction to Probability Theory and its Applications*, Vol. I, Third edition, John Wiley and Sons, Inc., New York-London-Sydney 1968 xviii+509 pp.

[35] Feynmann, R., Leighton, R. and Sands, M., *Feynmann's Lectures in Physics*, Vol. I, II and III, Addison-Wesley Pub Co, 1988.

[36] Finney, R. and Thomas, G., *Calculuc and Analytic Geometry*, Ninth edition, Addison-Wesley Pub Co., 1996.

[37] Fleming, Wendell, *Functions of Several Variables*, Second edition, Undergraduate Texts in Mathematics. Springer-Verlag, New York-Heidelberg, 1987. xi+411 pp.

[38] Folland, Gerald B., *Fourier Analysis and Its Applications*, The Wadsworth and Brooks/Cole Mathematics Series, Wadsworth and Brooks/Cole Advanced Books and Software, Pacific Grove, CA, 1992. 444 pp.

[39] Folland, Gerald B., *Introduction to Partial Differential Equations*, Second edition, Princeton University Press, Princeton, NJ, 1995. 352 pp.

[40] Folland, Gerald B., *Real analysis: Modern Techniques and their Application*, Second edition. Pure and Applied Mathematics. A Wiley-Interscience Publication, John Wiley and Sons, Inc., New York, 1999. xvi+386 pp.

[41] Fraleigh, John B., *A First Course in Abstract Algebra*, Sixth edition Addison-Wesley Pub Co. 1998, 576 pp.

[42] Fulton, W. and Harris, J., *Representation Theory: A First Course*, Graduate Texts in Mathematics, 129, Springer-Verlag, New York, 1991.

[43] Gallian, J., *Contemporary Abstract Algebra*, Fouth edition, Houghton Mifflin College, 1998. 583 pp.

[44] Gans, David, *An Introduction to Non-Euclidean Geometry*, Academic Press, New York-London, 1973. xii+274 pp.

[45] Garling, D. J. H., *A Course in Galois Theory*, Cambridge University Press, Cambridge-New York, 1987. 176 pp.

[46] Goldstern, Martin and Judah, Haim, *The Incompleteness Phenomenon*, A K Peters, Ltd., Natick, MA, 1998. 264 pp.

[47] Graham, Ronald L., Knuth, Donald E.and Patashnik, Oren, *Concrete mathematics: A Foundation for Computer Science*, Second edition, Addison-Wesley Publishing Company, Reading, MA, 1994. xiv+657 pp.

[48] Gray, Alfred, *Modern Differential Geometry of Curves and Surfaces with Mathematica*, Second edition, CRC Press, Boca Raton, FL, 1997. 1088 pp.

[49] Greene, Robert E.and Krantz, Steven G., *Function Theory of One Complex Variable*, Pure and Applied Mathematics. A Wiley-Interscience Publication, John Wiley and Sons, Inc., New York, 1997. xiv+496 pp.

[50] Grimmett, G. R.and Stirzaker, D. R., *Probability and Random Processes*, Second edition, The Clarendon Press, Oxford University Press, New York, 1992. 600 pp.

[51] Halliday, D., Resnick, R and Walker, J., *Fundamentals of Physics*, Fifth edition, John Wiley and Sons, 5th edition, 1996, 1142 pp/

[52] Halmos, Paul R., *Finite-Dimensional Vector Spaces*, Reprinting of the 1958 second edition, Undergraduate Texts in Mathematics, Springer-Verlag, New York-Heidelberg, 1993. viii+200 pp.

[53] Halmos, Paul R. *Naive Set Theory*, Reprinting of the 1960 edition, Undergraduate Texts in Mathematics, Springer-Verlag, New York-Heidelberg, 1974. vii+104 pp.

[54] Halmos, Paul R., *Measure Theory*, Graduate Texts in Mathematics, 18, Springer-Verlag, New York, 1976, 305 pp.

[55] Hartshorne, Robin, *Geometry: Euclid and Beyond*, Undergraduate Texts in Mathematics, Springer-Verlag, New York, 2000. xii+526 pp.

[56] David Henderson, *Differential Geometry: A Geometric Introduction*, Prentice Hall, 1998. 250 pp.

[57] Herstein, I., *Topics in Algebra*, Second edition, John Wiley & Sons, 1975.

[58] Hilbert, D. and Cohn-Vossen, S., *Geometry and the Imagination*, AMS Chelsea, 1999. 357 pp.

[59] Hill, Victor E., IV, *Groups and Characters*, Chapman and Hall/CRC, Boca Raton, FL, 1999.256 pp.

[60] Hintikka, Jaakko, *The Principles of Mathematics Revisited*, With an appendix by Gabriel Sandu, Cambridge University Press, Cambridge, 1998. 302 pp.

[61] Hofstadter, Douglas R., *Gödel, Escher, Bach: An Eternal Golden Braid*, Basic Books, Inc., Publishers, New York, 1979. 777 pp.

[62] Howard, Paul and Rubin, Jean, *Consequences of the Axiom of Choice*, Mathematical Surveys and Monographs, 59, American Mathematical Society, Providence, RI, 1998. viii+432 pp.

[63] Hubbard, Barbara Burke, *The World According to Wavelets: The Story of a Mathematical Technique in the Making*, Second edition, A K Peters, Ltd., Wellesley, MA, 1998. 286 pp.

[64] Hubbard, J. and Hubbard, B., *Vector Calculus, Linear Algebra, and Differential Forms: A Unified Approach*, Prentice Hall, 1999. 687 pp.

[65] Hungerford, T., *Algebra*, Eighth edition, Graduate Texts in Mathematics, 73, Springer, 1997. 502 pp.

[66] Iserles, Arieh, *A First Course in the Numerical Analysis of Differential Equations*, Cambridge Texts in Applied Mathematics, Cambridge University Press, Cambridge, 1996. 396 pp.

[67] Jackson, Dunham, *Fourier Series and Orthogonal Polynomials*, Carus Monograph Series, no. 6, Mathematical Association of America, Oberlin, Ohio, 1941. xii+234 pp.

[68] Jacobson, N., Basic Algebra, Vol. I and II, Second edition, W.H. Freeman, 1985.

[69] John, Fritz, *Partial Differential Equations*, Reprint of the fourth edition. Applied Mathematical Sciences, 1, Springer-Verlag, New York, 1991. x+249 pp.

[70] Jones, Frank, *Lebesgue Integration on Euclidean Space*, Revised edition, Jones and Bartlett Publishers, Boston, MA, 2001. 608 pp.

[71] Kac, Mark, *Statistical Independence in Probability, Analysis and Number Theory*, The Carus Mathematical Monographs, No. 12, Mathematical Association of America, New York 1969 xiv+93 pp.

[72] Kelley, John L., *General Topology*, Graduate Texts in Mathematics, 27. Springer-Verlag, New York-Berlin, 1975. xiv+298 pp.

[73] Kline, Morris, *Mathematics and the Search for Knowledge*, Oxford University Press, New York, 1972. 1256 pp.

[74] Kobayashi, Shoshichi and Nomizu, Katsumi, *Foundations of Differential Geometry*, Vol. I, Wiley Classics Library. A Wiley-Interscience Publication, John Wiley and Sons, Inc., New York, 1996. xii+329.

[75] Kobayashi, Shoshichi and Nomizu, Katsumi, *Foundations of Differential Geometry*, Vol. II, Wiley Classics Library, A Wiley-Interscience Publication, John Wiley and Sons, Inc., New York, 1996. xvi+468 pp

[76] Kolmogorov, A. N.and Fomin, S. V., *Introductory Real Analysis*, Translated from the second Russian edition and edited by Richard A. Silverman, Dover Publications, Inc., New York, 1975. xii+403 pp.

[77] Krantz, Steven G., *Function Theory of Several Complex Variables*, Second edition, AMS Chelsea, 2001. 564 pp.

[78] Krantz, Steven G., *Complex Analysis: The Geometric Viewpoint*, Carus Mathematical Monographs, 23, Mathematical Association of America, Washington, DC, 1990. 210 pp.

[79] Lang, Serge, *Algebra*, Third edition, Addison-Wesley, 1993 , 904 pp.

[80] Lang, Serge, *Undergraduate Analysis*, Second edition, Undergraduate Texts in Mathematics, Springer-Verlag, 1997, 642 pp.

[81] Lang, Serge and Murrow, Gene, *Geometry*, Second edition, Springer-Verlag, 2000, 394 pp.

[82] Mac Lane, Saunders, *Mathematics, Form and Function*, Springer-Verlag, New York-Berlin, 1986. xi+476 pp.

[83] Marsden, Jerrold E. and Hoffman, Michael J., *Basic Complex Analysis*, Third edition, W. H. Freeman and Company, New York, 1999. 600 pp.

[84] McCleary, John, *Geometry from a Differentiable Viewpoint*, Cambridge University Press, Cambridge, 1995. 320 pp.

[85] Millman, Richard and Parker, George D., *Elements of Differential Geometry* , Prentice-Hall Inc., Englewood Cliffs, N. J., 1977. xiv+265 pp.

[86] Morgan, Frank, *Riemannian Geometry: A Beginner's Guide*, Second edition, A K Peters, Ltd., Wellesley, MA, 1998. 160 pp.

[87] Moschovakis, Yiannis N., *Notes on Set Theory*, Undergraduate Texts in Mathematics, Springer-Verlag, New York, 1994. xiv+272 pp.

[88] Munkres, James R., *Topology: A First Course*, Second edition, Prentice-Hall, Inc., Englewood Cliffs, N.J., 2000. 537 pp.

[89] Nagel, Ernest and Newman, James R., *Gödel's Proof*, New York University Press, New York 1960 ix+118 pp.

[90] Olver, P., *Applications of Lie Groups to Diferential Equations*, Second edition, Graduate Texts in Mathematics, 107, Springer-Verlag, New York, 1993.

[91] O'Neill, Barrett, *Elementary Differential Geometry*, Second edition, Academic Press, New York-London 1997. 448 pp.

[92] Palka, Bruce P., *An Introduction to Complex Function Theory*, Undergraduate Texts in Mathematics, Springer-Verlag, New York, 1991. xviii+559 pp.

[93] Polya, George, Tarjan, Robert E.and Woods, Donald R., *Notes on Introductory Combinatorics*, Progress in Computer Science, 4, Birkhauser Boston, Inc., Boston, Mass., 1990. v+192 pp

[94] Protter, Murray H. and Morrey, Charles B., Jr. *A First Course in Real Analysis*, Second edition. Undergraduate Texts in Mathematics, Springer-Verlag, New York, 1991. xviii+534 pp.

[95] Royden, H. L., *Real Analysis*, Third edition, Prentice-Hall, 1988. 434 pp.

[96] Rudin, Walter. *Principles of Mathematical Analysis*, Third edition, International Series in Pure and Applied Mathematics, McGraw-Hill Book Co., New York-Auckland-Dsseldorf, 1976. x+342 pp.

[97] Rudin, Walter, *Real and complex analysis*, Third edition, McGraw-Hill Book Co., New York, 1986. xiv+416 pp.

[98] Seeley, Robert T., *An Introduction to Fourier Series and Integrals*, W. A. Benjamin, Inc., New York-Amsterdam 1966 x+104 pp.

[99] Simmons, George, *Differential Equations With Applications and Historical Notes*, McGraw-Hill Higher Education, 1991, 640 pp.

[100] Smullyan, Raymond M., *Gödel's Incompleteness Theorems*, Oxford Logic Guides, 19, The Clarendon Press, Oxford University Press, New York, 1992. xvi+139 pp.

[101] Spiegel, M., *Schaum's Outline of Complex Variables*, McGraw-Hill, 1983.

[102] Spivak, M., *Calculus*, Third edition, Publish or Perish, 1994. 670 pp.

[103] Spivak, Michael, *Calculus on Manifolds: A Modern Approach to Classical Theorems of Advanced Calculus*, Westview Press, 1971. 160 pp.

[104] Spivak, Michael, *A Comprehensive Introduction to Differential Geometry*, Vol. I - V. Third edition, Publish or Perish, Inc., 1979.

[105] Stanley, Richard P., *Enumerative Combinatorics*, Vol. 1, With a foreword by Gian-Carlo Rota, Cambridge Studies in Advanced Mathematics, 49. Cambridge University Press, Cambridge, 1997. 337 pp.

[106] Sternberg, S., *Group Theory and Physics*, Cambridge University Press, Cambridge, 1995. 443 pp.

[107] Stewart, Ian, *Galois theory*, Second edition, Chapman and Hall, Ltd., London, 1990. xxx+202 pp.

[108] Stewart, J., *Calculus*, Brooks/Cole Pub Co, third edition, 1995, 1015 pp.

[109] Strang, G., *Linear Algebra and its Applications*, Third edition, Harcourt College, 1988. 505 pp.

[110] Strikwerda, John C.. *Finite Difference Schemes and Partial Differential Equations*, The Wadsworth and Brooks/Cole Mathematics Series, Wadsworth and Brooks/Cole Advanced Books and Software, Pacific Grove, CA, 1989. xii+386 pp.

[111] Thorpe, John A., *Elementary Topics in Differential Geometry*, Undergraduate Texts in Mathematics, Springer-Verlag, New York, 1994. xiv+267 pp.

[112] Trefethen, Lloyd and Bau, David, III, *Numerical Linear Algebra*, Society for Industrial and Applied Mathematics (SIAM), Philadelphia, PA, 1997. xii+361 pp.

[113] van der Waerden, B. L., *Algebra*, Vol 1, Based in part on lectures by E. Artin and E. Noether, Translated from the seventh German edition by Fred Blum and John R. Schulenberger, Springer-Verlag, New York, 1991. xiv+265 pp.

[114] van der Waerden, B. L., *Algebra*, Vol 2, Based in part on lectures by E. Artin and E. Noether, Translated from the fifth German edition by John R. Schulenberger, Springer-Verlag, New York, 1991. xii+284 pp.

[115] van Lint, J. H. and Wilson, R. M., *A Course in Combinatorics*, Second edition, Cambridge University Press, Cambridge, 2001. 550 pp.

[116] Zygmund, A., *Trigonometric Series*, Vol. I, II, Reprinting of the 1968 version of the second edition with Volumes I and II bound together, Cambridge University Press, Cambridge-New York-Melbourne, 1988. 768 pp.

Index